PREPARING FOR THE FUTURE OF ARTIFICIAL INTELLIGENCE

Executive Office of the President
National Science and Technology Council
Committee on Technology

October 2016

Printed in the U.S.A.
ISBN-13: 978-1544643137

About the National Science and Technology Council

The National Science and Technology Council (NSTC) is the principal means by which the Executive Branch coordinates science and technology policy across the diverse entities that make up the Federal research and development (R&D) enterprise. One of the NSTC's primary objectives is establishing clear national goals for Federal science and technology investments. The NSTC prepares R&D packages aimed at accomplishing multiple national goals. The NSTC's work is organized under five committees: Environment, Natural Resources, and Sustainability; Homeland and National Security; Science, Technology, Engineering, and Mathematics (STEM) Education; Science; and Technology. Each of these committees oversees subcommittees and working groups that are focused on different aspects of science and technology. More information is available at www.whitehouse.gov/ostp/nstc.

About the Office of Science and Technology Policy

The Office of Science and Technology Policy (OSTP) was established by the National Science and Technology Policy, Organization, and Priorities Act of 1976. OSTP's responsibilities include advising the President in policy formulation and budget development on questions in which science and technology are important elements; articulating the President's science and technology policy and programs; and fostering strong partnerships among Federal, state, and local governments, and the scientific communities in industry and academia. The Director of OSTP also serves as Assistant to the President for Science and Technology and manages the NSTC. More information is available at www.whitehouse.gov/ostp.

Acknowledgments

This document was developed through the contributions of staff from OSTP, other components of the Executive Office of the President, and other departments and agencies. A special thanks and appreciation to everyone who contributed.

EXECUTIVE OFFICE OF THE PRESIDENT
NATIONAL SCIENCE AND TECHNOLOGY COUNCIL
WASHINGTON, D.C. 20502

October 12, 2016

Dear colleagues:

Advances in Artificial Intelligence (AI) technology have opened up new markets and new opportunities for progress in critical areas such as health, education, energy, and the environment. In recent years, machines have surpassed humans in the performance of certain specific tasks, such as some aspects of image recognition. Experts forecast that rapid progress in the field of specialized artificial intelligence will continue. Although it is very unlikely that machines will exhibit broadly-applicable intelligence comparable to or exceeding that of humans in the next 20 years, it is to be expected that machines will reach and exceed human performance on more and more tasks.

As a contribution toward preparing the United States for a future in which AI plays a growing role, this report surveys the current state of AI, its existing and potential applications, and the questions that are raised for society and public policy by progress in AI. The report also makes recommendations for specific further actions by Federal agencies and other actors. A companion document lays out a strategic plan for Federally-funded research and development in AI. Additionally, in the coming months, the Administration will release a follow-on report exploring in greater depth the effect of AI-driven automation on jobs and the economy.

The report was developed by the NSTC's Subcommittee on Machine Learning and Artificial Intelligence, which was chartered in May 2016 to foster interagency coordination, to provide technical and policy advice on topics related to AI, and to monitor the development of AI technologies across industry, the research community, and the Federal Government. The report was reviewed by the NSTC Committee on Technology, which concurred with its contents. The report follows a series of public-outreach activities spearheaded by the White House Office of Science and Technology Policy (OSTP) in 2016, which included five public workshops co-hosted with universities and other associations that are referenced in this report.

OSTP also published a Request for Information (RFI) in June 2016, which received 161 responses. The submitted comments were published by OSTP on September 6, 2016. Consistent with the role of Big Data as an enabler of AI, this report builds on three previous Administration reports on Big Data referenced in this report.

In the coming years, AI will continue to contribute to economic growth and will be a valuable tool for improving the world, as long as industry, civil society, and government work together to develop the positive aspects of the technology, manage its risks and challenges, and ensure that everyone has the opportunity to help in building an AI-enhanced society and to participate in its benefits.

Sincerely,

John P. Holdren
Assistant to the President for Science and Technology
Director, Office of Science and Technology Policy

Megan Smith
U.S. Chief Technology Officer

National Science and Technology Council

Chair
John P. Holdren
Assistant to the President for Science and Technology and Director, Office of Science and Technology Policy

Staff
Afua Bruce
Executive Director
Office of Science and Technology Policy

Subcommittee on Machine Learning and Artificial Intelligence

Co-Chair
Ed Felten
Deputy U.S. Chief Technology Officer
Office of Science and Technology Policy

Executive Secretary
Terah Lyons
Policy Advisor to the U.S. Chief Technology Officer
Office of Science and Technology Policy

Co-Chair
Michael Garris
Senior Scientist
National Institute of Standards and Technology
U.S. Department of Commerce

The following Federal departments and agencies are represented on the Subcommittee on Machine Learning and Artificial Intelligence and through it, work together to monitor the state of the art in machine learning (ML) and AI (within the Federal Government, in the private sector, and internationally), to watch for the arrival of important technology milestones in the development of AI, to coordinate the use of and foster the sharing of knowledge and best practices about ML and AI by the Federal Government, and to consult in the development of Federal research and development priorities in AI:

Department of Commerce (Co-Chair)
Department of Defense
Department of Education
Department of Energy
Department of Health and Human Services
Department of Homeland Security
Department of Justice
Department of Labor
Department of State
Department of Transportation
Department of Treasury

Department of Veterans Affairs
United States Agency for International Development
Central Intelligence Agency
General Services Administration
National Science Foundation
National Security Agency
National Aeronautics and Space Administration
Office of the Director of National Intelligence
Social Security Administration

The following offices of the Executive Office of the President are also represented on the Subcommittee:

Council of Economic Advisers
Domestic Policy Council
Office of Management and Budget
Office of Science and Technology Policy (Co-Chair)

Office of the Vice President
National Economic Council
National Security Council

PREPARING FOR THE FUTURE OF ARTIFICIAL INTELLIGENCE

Contents

Executive Summary 1
Introduction 5
    A Brief History of AI 5
    What is Artificial Intelligence? 6
    The Current State of AI 7
Public Outreach and Development of this Report 12
Applications of AI for Public Good 13
AI in the Federal Government 15
AI and Regulation 17
    Case Study: Autonomous Vehicles and Aircraft 18
Research and Workforce 23
    Monitoring Progress in AI 23
    Federal Support for AI Research 25
    Workforce Development and Diversity 26
AI, Automation, and the Economy 29
Fairness, Safety, and Governance 30
    Justice, Fairness, and Accountability 30
    Safety and Control 32
Global Considerations and Security 35
    International Cooperation 35
    AI and Cybersecurity 36
    AI in Weapon Systems 37
Conclusion 39
Recommendations in this Report 40
Acronyms 43
References 45

PREPARING FOR THE FUTURE OF ARTIFICIAL INTELLIGENCE

Executive Summary

As a contribution toward preparing the United States for a future in which Artificial Intelligence (AI) plays a growing role, we survey the current state of AI, its existing and potential applications, and the questions that are raised for society and public policy by progress in AI. We also make recommendations for specific further actions by Federal agencies and other actors. A companion document called the *National Artificial Intelligence Research and Development Strategic Plan* lays out a strategic plan for Federally-funded research and development in AI.

Applications of AI for Public Good

One area of great optimism about AI and machine learning is their potential to improve people's lives by helping to solve some of the world's greatest challenges and inefficiencies. Many have compared the promise of AI to the transformative impacts of advancements in mobile computing. Public- and private-sector investments in basic and applied R&D on AI have already begun reaping major benefits to the public in fields as diverse as health care, transportation, the environment, criminal justice, and economic inclusion. The effectiveness of government itself is being increased as agencies build their capacity to use AI to carry out their missions more quickly, responsively, and efficiently.

AI and Regulation

AI has applications in many products, such as cars and aircraft, which are subject to regulation designed to protect the public from harm and ensure fairness in economic competition. How will the incorporation of AI into these products affect the relevant regulatory approaches? In general, the approach to regulation of AI-enabled products to protect public safety should be informed by assessment of the aspects of risk that the addition of AI may reduce alongside the aspects of risk that it may increase. If a risk falls within the bounds of an existing regulatory regime, moreover, the policy discussion should start by considering whether the existing regulations already adequately address the risk, or whether they need to be adapted to the addition of AI. Also, where regulatory responses to the addition of AI threaten to increase the cost of compliance, or slow the development or adoption of beneficial innovations, policymakers should consider how those responses could be adjusted to lower costs and barriers to innovation without adversely impacting safety or market fairness.

Currently relevant examples of the regulatory challenges that AI-enabled products present are found in the cases of automated vehicles (AVs, such as self-driving cars) and AI-equipped unmanned aircraft systems (UAS, or "drones"). In the long run, AVs will likely save many lives by reducing driver error and increasing personal mobility, and UAS will offer many economic benefits. Yet public safety must be protected as these technologies are tested and begin to mature. The Department of Transportation (DOT) is using an approach to evolving the relevant regulations that is based on building expertise in the Department, creating safe spaces and test beds for experimentation, and working with industry and civil society to evolve performance-based regulations that will enable more uses as evidence of safe operation accumulates.

Research and Workforce

Government also has an important role to play in the advancement of AI through research and development and the growth of a skilled, diverse workforce. A separate strategic plan for Federally-funded AI research and development is being released in conjunction with this report. The plan discusses the role of Federal R&D, identifies areas of opportunity, and recommends ways to coordinate R&D to maximize benefit and build a highly-trained workforce.

Given the strategic importance of AI, moreover, it is appropriate for the Federal Government to monitor developments in the field worldwide in order to get early warning of important changes arising elsewhere in case these require changes in U.S. policy.

The rapid growth of AI has dramatically increased the need for people with relevant skills to support and advance the field. An AI-enabled world demands a data-literate citizenry that is able to read, use, interpret, and communicate about data, and participate in policy debates about matters affected by AI. AI knowledge and education are increasingly emphasized in Federal Science, Technology, Engineering, and Mathematics (STEM) education programs. AI education is also a component of Computer Science for All, the President's initiative to empower all American students from kindergarten through high school to learn computer science and be equipped with the computational thinking skills they need in a technology-driven world.

Economic Impacts of AI

AI's central economic effect in the short term will be the automation of tasks that could not be automated before. This will likely increase productivity and create wealth, but it may also affect particular types of jobs in different ways, reducing demand for certain skills that can be automated while increasing demand for other skills that are complementary to AI. Analysis by the White House Council of Economic Advisors (CEA) suggests that the negative effect of automation will be greatest on lower-wage jobs, and that there is a risk that AI-driven automation will increase the wage gap between less-educated and more-educated workers, potentially increasing economic inequality. Public policy can address these risks, ensuring that workers are retrained and able to succeed in occupations that are complementary to, rather than competing with, automation. Public policy can also ensure that the economic benefits created by AI are shared broadly, and assure that AI responsibly ushers in a new age in the global economy.

Fairness, Safety, and Governance

As AI technologies move toward broader deployment, technical experts, policy analysts, and ethicists have raised concerns about unintended consequences of widespread adoption. Use of AI to make consequential decisions about people, often replacing decisions made by human-driven bureaucratic processes, leads to concerns about how to ensure justice, fairness, and accountability—the same concerns voiced previously in the Administration's *Big Data: Seizing Opportunities, Preserving Values* report of 2014,[1] as well as the Report to the President on *Big Data and Privacy: A Technological Perspective* published by the President's Council of Advisors on Science and Technology in 2014.[2] Transparency concerns focus not only on the data and algorithms involved, but also on the potential to have some form of explanation for any AI-based determination. Yet AI experts have cautioned that there are inherent challenges in trying to understand and predict the behavior of advanced AI systems.

Use of AI to control physical-world equipment leads to concerns about safety, especially as systems are exposed to the full complexity of the human environment. A major challenge in AI safety is building systems that can safely transition from the "closed world" of the laboratory into the outside "open world" where unpredictable things can happen. Adapting gracefully to unforeseen situations is difficult yet necessary for safe operation. Experience in building other types of safety-critical systems and infrastructure, such as aircraft, power plants, bridges, and vehicles, has much to teach AI practitioners

---

[1] "Big Data: Seizing Opportunities, Preserving Values," *Executive Office of the President*, May 2014, https://www.whitehouse.gov/sites/default/files/docs/big_data_privacy_report_may_1_2014.pdf.

[2] The President's Council of Advisors on Science and Technology, "Report to the President: Big Data and Privacy: A Technological Perspective," *Executive Office of the President*, May 2014, https://www.whitehouse.gov/sites/default/files/microsites/ostp/PCAST/pcast_big_data_and_privacy_-_may_2014.pdf.

about verification and validation, how to build a safety case for a technology, how to manage risk, and how to communicate with stakeholders about risk.

At a technical level, the challenges of fairness and safety are related. In both cases, practitioners strive to avoid unintended behavior, and to generate the evidence needed to give stakeholders justified confidence that unintended failures are unlikely.

Ethical training for AI practitioners and students is a necessary part of the solution. Ideally, every student learning AI, computer science, or data science would be exposed to curriculum and discussion on related ethics and security topics. However, ethics alone is not sufficient. Ethics can help practitioners understand their responsibilities to all stakeholders, but ethical training should be augmented with technical tools and methods for putting good intentions into practice by doing the technical work needed to prevent unacceptable outcomes.

Global Considerations and Security

AI poses policy questions across a range of areas in international relations and security. AI has been a topic of interest in recent international discussions as countries, multilateral institutions, and other stakeholders have begun to access the benefits and challenges of AI. Dialogue and cooperation between these entities could help advance AI R&D and harness AI for good, while also addressing shared challenges.

Today's AI has important applications in cybersecurity, and is expected to play an increasing role for both defensive and offensive cyber measures. Currently, designing and operating secure systems requires significant time and attention from experts. Automating this expert work partially or entirely may increase security across a much broader range of systems and applications at dramatically lower cost, and could increase the agility of the Nation's cyber-defenses. Using AI may help maintain the rapid response required to detect and react to the landscape of evolving threats.

Challenging issues are raised by the potential use of AI in weapon systems. The United States has incorporated autonomy in certain weapon systems for decades, allowing for greater precision in the use of weapons and safer, more humane military operations. Nonetheless, moving away from direct human control of weapon systems involves some risks and can raise legal and ethical questions.

The key to incorporating autonomous and semi-autonomous weapon systems into American defense planning is to ensure that U.S. Government entities are always acting in accordance with international humanitarian law, taking appropriate steps to control proliferation, and working with partners and Allies to develop standards related to the development and use of such weapon systems. The United States has actively participated in ongoing international discussion on Lethal Autonomous Weapon Systems, and anticipates continued robust international discussion of these potential weapon systems. Agencies across the U.S. Government are working to develop a single, government-wide policy, consistent with international humanitarian law, on autonomous and semi-autonomous weapons.

Preparing for the Future

AI holds the potential to be a major driver of economic growth and social progress, if industry, civil society, government, and the public work together to support development of the technology with thoughtful attention to its potential and to managing its risks.

The U.S. Government has several roles to play. It can convene conversations about important issues and help to set the agenda for public debate. It can monitor the safety and fairness of applications as they develop, and adapt regulatory frameworks to encourage innovation while protecting the public. It can provide public policy tools to ensure that disruption in the means and methods of work enabled by AI increases productivity while avoiding negative economic consequences for certain sectors of the workforce. It can support basic research and the application of AI to public good. It can support development of a skilled, diverse workforce. And government can use AI itself to serve the public faster, more effectively, and at lower cost. Many areas of public policy, from education and the economic safety net, to defense, environmental preservation, and criminal justice, will see new opportunities and new challenges driven by the continued progress of AI. The U.S. Government must continue to build its capacity to understand and adapt to these changes.

As the technology of AI continues to develop, practitioners must ensure that AI-enabled systems are governable; that they are open, transparent, and understandable; that they can work effectively with people; and that their operation will remain consistent with human values and aspirations. Researchers and practitioners have increased their attention to these challenges, and should continue to focus on them.

Developing and studying machine intelligence can help us better understand and appreciate our human intelligence. Used thoughtfully, AI can augment our intelligence, helping us chart a better and wiser path forward.

A full list of the recommendations made in this report is on page 40.

Introduction

Artificial Intelligence (AI) has the potential to help address some of the biggest challenges that society faces. Smart vehicles may save hundreds of thousands of lives every year worldwide, and increase mobility for the elderly and those with disabilities. Smart buildings may save energy and reduce carbon emissions. Precision medicine may extend life and increase quality of life. Smarter government may serve citizens more quickly and precisely, better protect those at risk, and save money. AI-enhanced education may help teachers give every child an education that opens doors to a secure and fulfilling life. These are just a few of the potential benefits if the technology is developed with an eye to its benefits and with careful consideration of its risks and challenges.

The United States has been at the forefront of foundational research in AI, primarily supported for most of the field's history by Federal research funding and work at government laboratories. The Federal Government's support for unclassified AI R&D is managed through the Networking and Information Technology Research and Development (NITRD) program, and supported primarily by the Defense Advanced Research Projects Agency (DARPA), the National Science Foundation (NSF), the National Institutes of Health (NIH), the Office of Naval Research (ONR), and the Intelligence Advanced Research Projects Activity (IARPA). Major national research efforts such as the National Strategic Computing Initiative, the Big Data Initiative, and the Brain Research through Advancing Innovative Neurotechnologies (BRAIN) Initiative also contribute indirectly to the progress of AI research. The current and projected benefits of AI technology are large, adding to the Nation's economic vitality and to the productivity and well-being of its people. A companion document lays out a strategic plan for Federally-funded research and development in AI.

As a contribution toward preparing the United States for a future in which AI plays a growing role, we survey the current state of AI, its existing and potential applications, and the questions that progress in AI raise for society and public policy. We also make recommendations for specific further actions by Federal agencies and other actors.

A Brief History of AI

Endowing computers with human-like intelligence has been a dream of computer experts since the dawn of electronic computing. Although the term "Artificial Intelligence" was not coined until 1956, the roots of the field go back to at least the 1940s,[3] and the idea of AI was crystalized in Alan Turing's famous 1950 paper, "Computing Machinery and Intelligence." Turing's paper posed the question: "Can machines think?" It also proposed a test for answering that question,[4] and raised the possibility that a machine might be programmed to learn from experience much as a young child does.

In the ensuing decades, the field of AI went through ups and downs as some AI research problems proved more difficult than anticipated and others proved insurmountable with the technologies of the time. It wasn't until the late 1990s that research progress in AI began to accelerate, as researchers focused more on sub-problems of AI and the application of AI to real-world problems such as image recognition and medical diagnosis. An early milestone was the 1997 victory of IBM's chess-playing computer Deep Blue

---

[3] See, e.g., Warren S. McCulloch and Walter H. Pitts, "A Logical Calculus of the Ideas Immanent in Nervous Activity," *Bulletin of Mathematical Biophysics*, 5:115-133, 1943.

[4] Restated in modern terms, the "Turing Test" puts a human judge in a text-based chat room with either another person or a computer. The human judge can interrogate the other party and carry on a conversation, and then the judge is asked to guess whether the other party is a person or a computer. If a computer can consistently fool human judges in this game, then the computer is deemed to be exhibiting intelligence.

over world champion Garry Kasparov. Other significant breakthroughs included DARPA's Cognitive Agent that Learns and Organizes (CALO), which led to Apple Inc.'s Siri; IBM's question-answering computer Watson's victory in the TV game show "Jeopardy!"; and the surprising success of self-driving cars in the DARPA Grand Challenge competitions in the 2000s.

The current wave of progress and enthusiasm for AI began around 2010, driven by three factors that built upon each other: the availability of *big data* from sources including e-commerce, businesses, social media, science, and government; which provided raw material for dramatically *improved machine learning approaches and algorithms;* which in turn relied on the capabilities of *more powerful computers.*[5] During this period, the pace of improvement surprised AI experts. For example, on a popular image recognition challenge[6] that has a 5 percent human error rate according to one error measure, the best AI result improved from a 26 percent error rate in 2011 to 3.5 percent in 2015.

Simultaneously, industry has been increasing its investment in AI. In 2016, Google Chief Executive Officer (CEO) Sundar Pichai said, "Machine learning [a subfield of AI] is a core, transformative way by which we're rethinking how we're doing everything. We are thoughtfully applying it across all our products, be it search, ads, YouTube, or Play. And we're in early days, but you will see us—in a systematic way—apply machine learning in all these areas."[7] This view of AI broadly impacting how software is created and delivered was widely shared by CEOs in the technology industry, including Ginni Rometty of IBM, who has said that her organization is betting the company on AI.[8]

What is Artificial Intelligence?

There is no single definition of AI that is universally accepted by practitioners. Some define AI loosely as a computerized system that exhibits behavior that is commonly thought of as requiring intelligence. Others define AI as a system capable of rationally solving complex problems or taking appropriate actions to achieve its goals in whatever real world circumstances it encounters.

Experts offer differing taxonomies of AI problems and solutions. A popular AI textbook[9] used the following taxonomy: (1) systems that think like humans (e.g., cognitive architectures and neural networks); (2) systems that act like humans (e.g., pass the Turing test via natural language processing; knowledge representation, automated reasoning, and learning), (3) systems that think rationally (e.g.,

---

[5] A more detailed history of AI is available in the Appendix of the AI 100 Report. Peter Stone, Rodney Brooks, Erik Brynjolfsson, Ryan Calo, Oren Etzioni, Greg Hager, Julia Hirschberg, Shivaram Kalyanakrishnan, Ece Kamar, Sarit Kraus, Kevin Leyton-Brown, David Parkes, William Press, AnnaLee Saxenian, Julie Shah, Milind Tambe, and Astro Teller, "Artificial Intelligence and Life in 2030," *One Hundred Year Study on Artificial Intelligence: Report of the 2015-2016 Study Panel*, Stanford University, Stanford, CA, September 2016, http://ai100.stanford.edu/2016-report.

[6] The ImageNet Large Scale Visual Recognition Challenge provides a set of photographic images and asks for an accurate description of what is depicted in each image. Statistics in the text refer to the "classification error" metric in the "classification+localization with provided training data" task. See http://image-net.org/challenges/LSVRC/.

[7] Steven Levy, "How Google is Remaking Itself as a Machine Learning First Company," *Backchannel*, June 22, 2016, https://backchannel.com/how-google-is-remaking-itself-as-a-machine-learning-first-company-ada63defcb70.

[8] See, e.g., Andrew Nusca, "IBM's CEO Thinks Every Digital Business Will Become a Cognitive Computing Business," *Fortune*, June 1 2016. ("[IBM] CEO Ginni Rometty is optimistic that the company's wager on 'cognitive computing,' the term it uses for applied artificial intelligence and machine learning technologies, is the biggest bet the company will make in its 105-year history.")

[9] Stuart Russell and Peter Norvig, *Artificial Intelligence: A Modern Approach (3rd Edition)* (Essex, England: Pearson, 2009).

logic solvers, inference, and optimization); and (4) systems that act rationally (e.g., intelligent software agents and embodied robots that achieve goals via perception, planning, reasoning, learning, communicating, decision-making, and acting). Separately, venture capitalist Frank Chen broke down the problem space of AI into five general categories: logical reasoning, knowledge representation, planning and navigation, natural language processing, and perception.[10] And AI researcher Pedro Domingos ascribed AI researchers to five "tribes" based on the methods they use: "symbolists" use logical reasoning based on abstract symbols, "connectionists" build structures inspired by the human brain; "evolutionaries" use methods inspired by Darwinian evolution; "Bayesians" use probabilistic inference; and "analogizers" extrapolate from similar cases seen previously.[11]

This diversity of AI problems and solutions, and the foundation of AI in human evaluation of the performance and accuracy of algorithms, makes it difficult to clearly define a bright-line distinction between what constitutes AI and what does not. For example, many techniques used to analyze large volumes of data were developed by AI researchers and are now identified as "Big Data" algorithms and systems. In some cases, opinion may shift, meaning that a problem is considered as requiring AI before it has been solved, but once a solution is well known it is considered routine data processing. Although the boundaries of AI can be uncertain and have tended to shift over time, what is important is that a core objective of AI research and applications over the years has been to automate or replicate intelligent behavior.

The Current State of AI

Remarkable progress has been made on what is known as *Narrow AI,* which addresses specific application areas such as playing strategic games, language translation, self-driving vehicles, and image recognition.[12] Narrow AI underpins many commercial services such as trip planning, shopper recommendation systems, and ad targeting, and is finding important applications in medical diagnosis, education, and scientific research. These have all had significant societal benefits and have contributed to the economic vitality of the Nation.[13]

General AI (sometimes called Artificial General Intelligence, or AGI) refers to a notional future AI system that exhibits apparently intelligent behavior at least as advanced as a person across the full range of cognitive tasks. A broad chasm seems to separate today's Narrow AI from the much more difficult challenge of General AI. Attempts to reach General AI by expanding Narrow AI solutions have made little headway over many decades of research. The current consensus of the private-sector expert community, with which the NSTC Committee on Technology concurs, is that General AI will not be achieved for at least decades.[14]

---

[10] Frank Chen, "AI, Deep Learning, and Machine Learning: A Primer," *Andreessen Horowitz*, June 10, 2016, http://a16z.com/2016/06/10/ai-deep-learning-machines.

[11] Pedro Domingos, *The Master Algorithm: How the Quest for the Ultimate Learning Machine Will Remake Our World* (New York, New York: Basic Books, 2015).

[12] Narrow AI is not a single technical approach, but rather a set of discrete problems whose solutions rely on a toolkit of AI methods along with some problem-specific algorithms. The diversity of Narrow AI problems and solutions, and the apparent need to develop specific methods for each Narrow AI application, has made it infeasible to "generalize" a single Narrow AI solution to produce intelligent behavior of general applicability.

[13] Mike Purdy and Paul Daugherty, "Why Artificial Intelligence is the Future of Growth," *Accenture*, 2016, https://www.accenture.com/us-en/_acnmedia/PDF-33/Accenture-Why-AI-is-the-Future-of-Growth.pdf.

[14] Expert opinion on the expected arrival date of AGI ranges from 2030 to centuries from now. There is a long history of excessive optimism about AI. For example, AI pioneer Herb Simon predicted in 1957 that computers

People have long speculated on the implications of computers becoming more intelligent than humans. Some predict that a sufficiently intelligent AI could be tasked with developing even better, more intelligent systems, and that these in turn could be used to create systems with yet greater intelligence, and so on, leading in principle to an "intelligence explosion" or "singularity" in which machines quickly race far ahead of humans in intelligence.[15]

In a dystopian vision of this process, these *super-intelligent* machines would exceed the ability of humanity to understand or control. If computers could exert control over many critical systems, the result could be havoc, with humans no longer in control of their destiny at best and extinct at worst. This scenario has long been the subject of science fiction stories, and recent pronouncements from some influential industry leaders have highlighted these fears.

A more positive view of the future held by many researchers sees instead the development of intelligent systems that work well as helpers, assistants, trainers, and teammates of humans, and are designed to operate safely and ethically.

The NSTC Committee on Technology's assessment is that long-term concerns about super-intelligent General AI should have little impact on current policy. The policies the Federal Government should adopt in the near-to-medium term if these fears are justified are almost exactly the same policies the Federal Government should adopt if they are not justified. The best way to build capacity for addressing the longer-term speculative risks is to attack the less extreme risks already seen today, such as current security, privacy, and safety risks, while investing in research on longer-term capabilities and how their challenges might be managed. Additionally, as research and applications in the field continue to mature, practitioners of AI in government and business should approach advances with appropriate consideration of the long-term societal and ethical questions – in additional to just the technical questions – that such advances portend. Although prudence dictates some attention to the possibility that harmful super-intelligence might someday become possible, these concerns should not be the main driver of public policy for AI.

Machine Learning

Machine learning is one of the most important technical approaches to AI and the basis of many recent advances and commercial applications of AI. Modern machine learning is a statistical process that starts with a body of data and tries to derive a rule or procedure that explains the data or can predict future data. This approach—learning from data—contrasts with the older "expert system" approach to AI, in which programmers sit down with human domain experts to learn the rules and criteria used to make decisions, and translate those rules into software code. An expert system aims to emulate the principles used by human experts, whereas machine learning relies on statistical methods to find a decision procedure that works well in practice.

An advantage of machine learning is that it can be used even in cases where it is infeasible or difficult to write down explicit rules to solve a problem. For example, a company that runs an online service might use machine learning to detect user log-in attempts that are fraudulent. The company might start with a large data set of past login attempts, with each attempt labeled as fraudulent or not using the benefit of

---

would outplay humans at chess within a decade, an outcome that required 40 years to occur. Early predictions about automated language translation also proved wildly optimistic, with the technology only becoming usable (and by no means fully fluent) in the last several years. It is tempting but incorrect to extrapolate from the ability to solve one particular task to imagine machines with a much broader and deeper range of capabilities and to overlook the huge gap between narrow task-oriented performance and the type of general intelligence that people exhibit.

[15] It is far from certain that this sort of explosive growth in intelligence is likely, or even possible. Another plausible extrapolation from current knowledge is that machine intelligence will continue to increase gradually even after surpassing human intelligence.

hindsight. Based on this data set, the company could use machine learning to derive a rule to apply to future login attempts that predicts which attempts are more likely to be fraudulent and should be subjected to extra security measures. In a sense, machine learning is not an algorithm for solving a specific problem, but rather a more general approach to finding solutions for many different problems, given data about them.

To apply machine learning, a practitioner starts with a historical data set, which the practitioner divides into a *training set* and a *test set*. The practitioner chooses a *model*, or mathematical structure that characterizes a range of possible decision-making rules with adjustable *parameters*. A common analogy is that the model is a "box" that applies a rule, and the parameters are adjustable knobs on the front of the box that control how the box operates. In practice, a model might have many millions of parameters.

The practitioner also defines an *objective function* used to evaluate the desirability of the outcome that results from a particular choice of parameters. The objective function will typically contain parts that reward the model for closely matching the training set, as well as parts that reward the use of simpler rules.

Training the model is the process of adjusting the parameters to maximize the objective function. Training is the difficult technical step in machine learning. A model with millions of parameters will have astronomically more possible outcomes than any algorithm could ever hope to try, so successful training algorithms have to be clever in how they explore the space of parameter settings so as to find very good settings with a feasible level of computational effort.

Once a model has been trained, the practitioner can use the test set to evaluate the accuracy and effectiveness of the model. The goal of machine learning is to create a trained model that will *generalize*—it will be accurate not only on examples in the training set, but also on future cases that it has never seen before. While many of these models can achieve better-than-human performance on narrow tasks such as image labeling, even the best models can fail in unpredictable ways. For example, for many image labeling models it is possible to create images that clearly appear to be random noise to a human but will be falsely labeled as a specific object with high confidence by a trained model.[16]

Another challenge in using machine learning is that it is typically not possible to extract or generate a straightforward explanation for why a particular trained model is effective. Because trained models have a very large number of adjustable parameters—often hundreds of millions or more—training may yield a model that "works," in the sense of matching the data, but is not necessarily the simplest model that works. In human decision-making, any opacity in the process is typically due to not having enough information about why a decision was reached, because the decider may be unable to articulate why the decision "felt right." With machine learning, everything about the decision procedure is known with mathematical precision, but there may be simply too much information to interpret clearly.

Deep Learning

In recent years, some of the most impressive advancements in machine learning have been in the subfield of deep learning, also known as deep network learning. Deep learning uses structures loosely inspired by the human brain, consisting of a set of units (or "neurons"). Each unit combines a set of input values to produce an output value, which in turn is passed on to other neurons downstream. For example, in an image recognition application, a first layer of units might combine the raw data of the image to recognize simple patterns in the image; a second layer of units might combine the results of the first layer to recognize patterns-of-patterns; a third layer might combine the results of the second layer; and so on.

---

[16] See, e.g., Ian J. Goodfellow, Jonathon Shlens, and Christian Szegedy, "Explaining and Harnessing Adversarial Examples," http://arxiv.org/pdf/1412.6572.pdf.

Deep learning networks typically use many layers—sometimes more than 100— and often use a large number of units at each layer, to enable the recognition of extremely complex, precise patterns in data.

In recent years, new theories of how to construct and train deep networks have emerged, as have larger, faster computer systems, enabling the use of much larger deep learning networks. The dramatic success of these very large networks at many machine learning tasks has come as a surprise to some experts, and is the main cause of the current wave of enthusiasm for machine learning among AI researchers and practitioners.

Autonomy and Automation

AI is often applied to systems that can control physical actuators or trigger online actions. When AI comes into contact with the everyday world, issues of autonomy, automation, and human-machine teaming arise.

Autonomy refers to the ability of a system to operate and adapt to changing circumstances with reduced or without human control. For example, an autonomous car could drive itself to its destination. Despite the focus in much of the literature on cars and aircraft, autonomy is a much broader concept that includes scenarios such as automated financial trading and automated content curation systems. Autonomy also includes systems that can diagnose and repair faults in their own operation, such as identifying and fixing security vulnerabilities.

Automation occurs when a machine does work that might previously have been done by a person.[17] The term relates to both physical work and mental or cognitive work that might be replaced by AI. Automation, and its impact on employment, have been significant social and economic phenomena since at least the Industrial Revolution. It is widely accepted that AI will automate some jobs, but there is more debate about whether this is just the next chapter in the history of automation or whether AI will affect the economy differently than past waves of automation have previously.

Human-Machine Teaming

In contrast to automation, where a machine substitutes for human work, in some cases a machine will complement human work. This may happen as a side-effect of AI development, or a system might be developed specifically with the goal of creating a human-machine team. Systems that aim to complement human cognitive capabilities are sometimes referred to as intelligence augmentation.

In many applications, a human-machine team can be more effective than either one alone, using the strengths of one to compensate for the weaknesses of the other. One example is in chess playing, where a weaker computer can often beat a stronger computer player, if the weaker computer is given a human teammate—this is true even though top computers are much stronger players than any human.[18] Another example is in radiology. In one recent study, given images of lymph node cells, and asked to determine

---

[17] Different definitions of "automation" are used in different settings. The definition used in the main text, involving the substitution of machine labor for human labor, is commonly used in economics. Another definition is used in the systems analysis setting in the Department of Defense (DoD): Automation means that the system functions with little or no human operator involvement. However the system performance is limited to the specific pre-programmed actions it has been designed to execute. Once the system is initiated by a human operator, it executes its task according to those instructions and subroutines, which have been tested and validated. Typically these are well-defined tasks that have predetermined responses, i.e., rule-based responses in reasonably well-known and structured environments.

[18] Garry Kasparov, "The Chess Master and the Computer," *New York Review of Books*, February 11, 2010. http://www.nybooks.com/articles/2010/02/11/the-chess-master-and-the-computer.

whether or not the cells contained cancer, an AI-based approach had a 7.5 percent error rate, where a human pathologist had a 3.5 percent error rate; a combined approach, using both AI and human input, lowered the error rate to 0.5 percent, representing an 85 percent reduction in error.[19]

---

[19] Dayong Wang, Aditya Khosla, Rishab Gargeya, Humayun Irshad, Andrew H. Beck, "Deep Learning for Identifying Metastatic Breast Cancer," June 18, 2016, https://arxiv.org/pdf/1606.05718v1.pdf.

Public Outreach and Development of this Report

This report was developed by the NSTC's Subcommittee on Machine Learning and Artificial Intelligence, which was chartered in May 2016 to foster interagency coordination and provide technical and policy advice on topics related to AI, and to monitor the development of AI technologies across industry, the research community, and the Federal Government. The report follows a series of public outreach activities led by OSTP, designed to allow government officials thinking about these topics to learn from experts and from the public. This public outreach on AI included five co-hosted public workshops, and a public Request for Information (RFI). The public workshops were:

- AI, Law, and Governance (May 24, in Seattle, co-hosted by OSTP, the National Economic Council (NEC), and the University of Washington);
- AI for Social Good (June 7, in Washington DC, co-hosted by OSTP, the Association for the Advancement of AI (AAAI) and the Computing Community Consortium (CCC));
- Future of AI: Emerging Topics and Societal Benefit at the Global Entrepreneurship Summit (June 23, in Palo Alto, co-hosted by OSTP and Stanford University);
- AI Technology, Safety, and Control (June 28, in Pittsburgh, co-hosted by OSTP and Carnegie Mellon University); and
- Social and Economic Impacts of AI (July 7, in New York, co-hosted by OSTP, NEC, and New York University).

In conjunction with each of the five workshops, the private-sector co-hosts organized separate meetings or conference sessions which government staff attended. Total in-person attendance at the public events was more than 2,000 people, in addition to international online streaming audiences, which included more than 3,500 people for the Washington, DC workshop livestream alone.

OSTP also published a Request for Information (RFI) seeking public comment on the topics of the workshops. The RFI closed on July 22, 2016 and received 161 responses. Comments submitted in response to the public RFI were published by OSTP on September 6, 2016.[20]

---

[20] Ed Felten and Terah Lyons, "Public Input and Next Steps on the Future of Artificial Intelligence," *Medium*, September 6 2016, https://medium.com/@USCTO/public-input-and-next-steps-on-the-future-of-artificial-intelligence-458b82059fc3.

Applications of AI for Public Good

One area of great optimism about AI and machine learning is their potential to improve people's lives by helping to solve some of the world's greatest challenges and inefficiencies. The promise of AI has been compared to the transformative impacts of advances in mobile computing.[21] Public- and private-sector investments in basic and applied R&D on AI have already begun reaping major benefits for the public in fields as diverse as health care, transportation, the environment, criminal justice, and economic inclusion.[22]

At Walter Reed Medical Center, the Department of Veteran Affairs is using AI to better predict medical complications and improve treatment of severe combat wounds, leading to better patient outcomes, faster healing, and lower costs.[23] The same general approach—predicting complications to enable preventive treatment—has also reduced hospital-acquired infections at Johns Hopkins University.[24] Given the current transition to electronic health records, predictive analysis of health data may play a key role across many health domains like precision medicine and cancer research.

In transportation, AI-enabled smarter traffic management applications are reducing wait times, energy use, and emissions by as much as 25 percent in some places.[25] Cities are now beginning to leverage the type of responsive dispatching and routing used by ride-hailing services, and linking it with scheduling and tracking software for public transportation to provide just-in-time access to public transportation that can often be faster, cheaper and, in many cases, more accessible to the public.

Some researchers are leveraging AI to improve animal migration tracking by using AI image classification software to analyze tourist photos from public social media sites. The software can identify individual animals in the photos and build a database of their migration using the data and location stamps on the photos. At OSTP's AI for Social Good workshop, researchers talked about building some of the largest available datasets to-date on the populations and migrations of whales and large African animals, and about launching a project to track "The Internet of Turtles" to gain new insights about sea life.[26] Other speakers described uses of AI to optimize the patrol strategy of anti-poaching agents, and to design habitat preservation strategies to maximize the genetic diversity of endangered populations.

---

[21] Frank Chen, "AI, Deep Learning, and Machine Learning: A Primer," *Andreessen Horowitz*, June 10, 2016, http://a16z.com/2016/06/10/ai-deep-learning-machines.

[22] The potential benefits of increasing access to digital technologies are detailed further in the World Bank Group's Digital Dividends report. ("World Development Report 2016: Digital Dividends," *The World Bank Group*, 2016, http://documents.worldbank.org/curated/en/896971468194972881/pdf/102725-PUB-Replacement-PUBLIC.pdf.)

[23] Eric Elster, "Surgical Critical Care Initiative: Bringing Precision Medicine to the Critically Ill," presentation at AI for Social Good workshop, Washington, DC, June 7, 2016, http://cra.org/ccc/wp-content/uploads/sites/2/2016/06/Eric-Elster-AI-slides-min.pdf.

[24] Katharine E. Henry, David N. Hager, Peter J. Pronovost, and Suchi Saria, "A targeted real-time early warning score (TREWScore) for septic shock," *Science Translational Medicine* 7, no. 299 (2015): 299ra122-299ra122.

[25] Stephen F. Smith, "Smart Infrastructure for Urban Mobility," presentation at AI for Social Good workshop, Washington, DC, June 7, 2016, http://cra.org/ccc/wp-content/uploads/sites/2/2016/06/Stephen-Smith-AI-slides.pdf.

[26] Aimee Leslie, Christine Hof, Diego Amorocho, Tanya Berger-Wolf, Jason Holmberg, Chuck Stewart, Stephen G. Dunbar, and Claire Jea,, "The Internet of Turtles," April 12, 2016, https://www.researchgate.net/publication/301202821_The_Internet_of_Turtles.

Autonomous sailboats and watercraft are already patrolling the oceans carrying sophisticated sensor instruments, collecting data on changes in Arctic ice and sensitive ocean ecosystems in operations that would be too expensive or dangerous for crewed vessels. Autonomous watercraft may be much cheaper to operate than manned ships, and may someday be used for enhanced weather prediction, climate monitoring, or policing illegal fishing.[27]

AI also has the potential to improve aspects of the criminal justice system, including crime reporting, policing, bail, sentencing, and parole decisions. The Administration is exploring how AI can responsibly benefit current initiatives such as Data Driven Justice and the Police Data Initiative that seek to provide law enforcement and the public with data that can better inform decision-making in the criminal justice system, while also taking care to minimize the possibility that AI might introduce bias or inaccuracies due to deficiencies in the available data.

Several U.S. academic institutions have launched initiatives to use AI to tackle economic and social challenges. For example, the University of Chicago created an academic program that uses data science and AI to address public challenges such as unemployment and school dropouts.[28] The University of Southern California launched the Center for Artificial Intelligence in Society, an institute dedicated to studying how computational game theory, machine learning, automated planning and multi-agent reasoning techniques can help to solve socially relevant problems like homelessness. Meanwhile, researchers at Stanford University are using machine learning in efforts to address global poverty by using AI to analyze satellite images of likely poverty zones to identify where help is needed most.[29]

Many uses of AI for public good rely on the availability of data that can be used to train machine learning models and test the performance of AI systems. Agencies and organizations with data that can be released without implicating personal privacy or trade secrets can help to enable the development of AI by making those data available to researchers. Standardizing data schemas and formats can reduce the cost and difficulty of making new data sets useful.

> *Recommendation 1: Private and public institutions are encouraged to examine whether and how they can responsibly leverage AI and machine learning in ways that will benefit society. Social justice and public policy institutions that do not typically engage with advanced technologies and data science in their work should consider partnerships with AI researchers and practitioners that can help apply AI tactics to the broad social problems these institutions already address in other ways.*
>
> *Recommendation 2: Federal agencies should prioritize open training data and open data standards in AI. The government should emphasize the release of datasets that enable the use of AI to address social challenges. Potential steps may include developing an "Open Data for AI" initiative with the objective of releasing a significant number of government data sets to accelerate AI research and galvanize the use of open data standards and best practices across government, academia, and the private sector.*

---

[27] John Markoff, "No Sailors Needed: Robot Sailboats Scout the Oceans for Data," *The New York Times*, September 4, 2016.

[28] "Data Science for Social Good," *University of Chicago, https://dssg.uchicago.edu/.*

[29] Neal Jean, Marshall Burke, Michael Xie, W. Matthew Davis, David B. Lobell, and Stefano Ermon. "Combining satellite imagery and machine learning to predict poverty." *Science* 353, no. 6301 (2016): 790-794.

AI in the Federal Government

The Administration is working to develop policies and internal practices that will maximize the economic and societal benefits of AI and promote innovation. These policies and practices may include:

- investing in basic and applied research and development (R&D);
- serving as an early customer for AI technologies and their applications;
- supporting pilot projects and creating testbeds in real-world settings;
- making data sets available to the public;
- sponsoring incentive prizes;
- identifying and pursuing Grand Challenges to set ambitious but achievable goals for AI;
- funding rigorous evaluations of AI applications to measure their impact and cost-effectiveness; and
- creating a policy, legal, and regulatory environment that allows innovation to flourish while protecting the public from harm.

Using AI in Government to Improve Services and Benefit the American People

One challenge in using AI to improve services is that the Federal Government's capacity to foster and harness innovation in order to better serve the country varies widely across agencies. Some agencies are more focused on innovation, particularly those agencies with large R&D budgets, a workforce that includes many scientists and engineers, a culture of innovation and experimentation, and strong ongoing collaborations with private-sector innovators. Many also have organizations that are specifically tasked with supporting high-risk, high-return research (e.g., the advanced research projects agencies in the Departments of Defense and Energy, as well as the Intelligence Community), and fund R&D across the full range from basic research to advanced development. Other agencies like the NSF have research and development as their primary mission.

But some agencies, particularly those charged with reducing poverty and increasing economic and social mobility, have more modest levels of relevant capabilities, resources, and expertise.[30] For example, while the National Institutes of Health (NIH) has an R&D budget of more than $30 billion, the Department of Labor's R&D budget is only $14 million. This limits the Department of Labor's capacity to explore applications of AI, such as applying AI-based "digital tutor" technology to increase the skills and incomes of non-college educated workers.

DARPA's "Education Dominance" program serves as an example of AI's potential to fulfill and accelerate agency priorities. DARPA, intending to reduce from years to months the time required for new Navy recruits to become experts in technical skills, now sponsors the development of a digital tutor that uses AI to model the interaction between an expert and a novice. An evaluation of the digital tutor program concluded that Navy recruits using the digital tutor to become IT systems administrators frequently outperform Navy experts with 7-10 years of experience in both written tests of knowledge and real-world problem solving.[31]

Preliminary evidence based on digital tutor pilot projects also suggests that workers who have completed a training program that uses the digital tutor are more likely to get a high-tech job that dramatically

---

[30] Thomas Kalil, "A Broader Vision for Government Research," *Issues in Science and Technology*, 2003.

[31] "Winning the Education Future: The Role of ARPA-ED," *The U.S. Department of Education*, March 8 2011, https://www.whitehouse.gov/sites/default/files/microsites/ostp/arpa-ed-factsheet.pdf.

increases their incomes.[32] These wage increases appear to be much larger than the impacts of current workforce development programs.[33] Ideally, these results would be confirmed with independently conducted, randomized, controlled trials. Currently, the cost of developing digital tutors is high, and there is no repeatable methodology for developing effective digital tutors. Research that enables the emergence of an industry that uses AI approaches such as digital tutors could potentially help workers acquire in-demand skills.

> *Recommendation 3: The Federal Government should explore ways to improve the capacity of key agencies to apply AI to their missions. For example, Federal agencies should explore the potential to create DARPA-like organizations to support high-risk, high-reward AI research and its application, much as the Department of Education has done through its proposal to create an "ARPA-ED," to support R&D to determine whether AI and other technologies could significantly improve student learning outcomes.*
>
> *Recommendation 4: The NSTC MLAI subcommittee should develop a community of practice for AI practitioners across government. Agencies should work together to develop and share standards and best practices around the use of AI in government operations. Agencies should ensure that Federal employee training programs include relevant AI opportunities.*

---

[32] The President's Council of Advisors on Science and Technology, letter to the President, September 2014, https://www.whitehouse.gov/sites/default/files/microsites/ostp/PCAST/pcast_workforce_edit_report_sept_2014.pdf.

[33] J.D. Fletcher, "Digital Tutoring in Information Systems Technology for Veterans: Data Report," *The Institute for Defense Analysis,* September 2014.

AI and Regulation

AI has applications in many products, such as cars and aircraft, which are subject to regulation designed to protect the public from harm and ensure fairness in economic competition. How will the incorporation of AI into these products affect the relevant regulatory approaches? In general, the approach to regulation of AI-enabled products to protect public safety should be informed by assessment of the aspects of risk that the addition of AI may reduce, alongside the aspects of risk that it may increase. If a risk falls within the bounds of an existing regulatory regime, moreover, the policy discussion should start by considering whether the existing regulations already adequately address the risk, or whether they need to be adapted to the addition of AI. Also, where regulatory responses to the addition of AI threaten to increase the cost of compliance or slow the development or adoption of beneficial innovations, policymakers should consider how those responses could be adjusted to lower costs and barriers to innovation without adversely impacting safety or market fairness.

The general consensus of the RFI commenters was that broad regulation of AI research or practice would be inadvisable at this time.[34] Instead, commenters said that the goals and structure of existing regulations were sufficient, and commenters called for existing regulation to be adapted as necessary to account for the effects of AI. For example, commenters suggested that motor vehicle regulation should evolve to account for the anticipated arrival of autonomous vehicles, and that the necessary evolution could be carried out within the current structure of vehicle safety regulation. In doing so, agencies must remain mindful of the fundamental purposes and goals of regulation to safeguard the public good, while creating space for innovation and growth in AI.

Effective regulation of technologies such as AI requires agencies to have in-house technical expertise to help guide regulatory decision-making. The need for senior-level expert participation exists at regulating departments and agencies, and at all stages of the regulatory process. A range of personnel assignment and exchange models (e.g. hiring authorities) can be used to develop a Federal workforce with more diverse perspectives on the current state of technological development. One example of such an authority is the Intergovernmental Personnel Act (IPA) Mobility Program, which provides for the temporary assignment of personnel between the Federal Government and state and local governments, colleges and universities, Indian tribal governments, federally funded research and development centers, and other eligible organizations. If used strategically, the IPA program can help agencies meet their needs for hard-to-fill positions and increase their ability to hire candidates from diverse technical backgrounds. Federal employees serving in IPA assignments can serve as both recruiters and ambassadors for the Federal workforce. For example, agency staff sent to colleges and universities as instructors can inspire students to consider Federal employment. Likewise, programs that rotate employees through different jobs and sectors can help government employees gain knowledge and experience to inform regulation and policy, especially as it relates to emergent technologies like AI.

---

[34] Ed Felten and Terah Lyons, "Public Input and Next Steps on the Future of Artificial Intelligence."

> ***Recommendation 5:*** *Agencies should draw on appropriate technical expertise at the senior level when setting regulatory policy for AI-enabled products. Effective regulation of AI-enabled products requires collaboration between agency leadership, staff knowledgeable about the existing regulatory framework and regulatory practices generally, and technical experts with knowledge of AI. Agency leadership should take steps to recruit the necessary technical talent, or identify it in existing agency staff, and should ensure that there are sufficient technical "seats at the table" in regulatory policy discussions.*
>
> ***Recommendation 6:*** *Agencies should use the full range of personnel assignment and exchange models (e.g. hiring authorities) to foster a Federal workforce with more diverse perspectives on the current state of technology.*

Case Study: Autonomous Vehicles and Aircraft

A relevant example of the regulatory challenges associated with an agency updating legacy regulations to account for new AI-based products is the work of the Department of Transportation (DOT) on automated vehicles and unmanned aircraft systems (UAS, or "drones"). Within DOT, automated cars are regulated by the National Highway Traffic Safety Administration (NHTSA) and aircraft are regulated by the Federal Aviation Administration (FAA).

The Promise of Autonomy

The application of AI to vehicles and aircraft has captured the public imagination. Today's new cars have AI-based driver assist features like self-parking and advanced cruise controls that keep a car in its lane and adjust speed based on surrounding vehicles. Experimental fully automated cars monitored by humans can already be seen driving on the roads. The consensus of experts is that automated surface vehicle technology will eventually be safer than human drivers and may someday prevent most of the tens of thousands of fatalities that occur annually on the Nation's roads and highways.

Automated vehicles also offer the possibility of greater mobility for the elderly and Americans with disabilities who may not be able to drive. First- and last-mile access to transit and other novel transportation approaches may provide communities isolated from essential services such as jobs, health care, and groceries unprecedented access to opportunity. A well-designed system of automated vehicles able to predict and avoid collisions may also significantly reduce transportation-related emissions and energy consumption. The Administration is taking steps to make this vision a reality, including the proposed $3.9 billion investment in the President's Fiscal Year (FY) 2017 Budget by the Department of Transportation in automated and connected vehicle research, development, and deployment efforts, to ensure that the United States maintains its lead in automated vehicle technologies.[35]

---

[35] "Secretary Foxx Unveils President Obama's FY17 Budget Proposal of Nearly $4 Billion for Automated Vehicles and Announces DOT Initiatives to Accelerate Vehicle Safety Innovations," *U.S. Department of Transportation*, January 14 2016, https://www.transportation.gov/briefing-room/secretary-foxx-unveils-president-obama%E2%80%99s-fy17-budget-proposal-nearly-4-billion.

Moving to the air, since the early 1990s, commercial UAS have operated on a limited basis in the National Airspace System (NAS).[36] Until recently, UAS mainly supported government operations, such as military and border security operations. But in recent years, potential applications have rapidly expanded to include aerial photography, surveying land and crops, monitoring forest fires, responding to disasters, and inspecting critical infrastructure. Several government agencies are already operating UAS to advance their missions, and thousands of Americans have obtained the necessary authority from the Federal Aviation Administration (FAA) for commercial UAS operations, a process accelerated under the FAA's "Small UAS Rule" that took effect in August 2016 and the FAA's Small UAS Aircraft Registration Service that launched in December 2015. The FAA estimates that the number of UAS registered for commercial use will exceed 600,000 by August 2017.[37]

One estimate of the economic impact of integrating of UAS into the airspace predicted more than $13.6 billion of economic value created by UAS in the first three years of integration, with sustainable growth predicted to follow.[38] A 2013 study from the Association for Unmanned Vehicle Systems International predicted that the commercial drone industry could generate more than $82 billion for the U.S. economy and create more than 100,000 new jobs over the next 10 years. Tax revenue to the states was predicted to increase by more than $482 million in the first decade after integration.[39]

Ensuring Safety

Realizing the potential benefits of these promising technologies requires that government take steps to ensure the safety of the airspace and roads, while continuing to foster a culture of innovation and growth. The United States has the safest and most complex aviation system in the world, and the public relies on FAA oversight to establish safety standards. Federal Motor Vehicle Safety Standards (FMVSS) place requirements on manufacturers to develop safe surface vehicles, and NHTSA has the authority to recall vehicles in the event of an unreasonable risk to safety. While there is considerable opportunity to reduce the fatalities on roads and highways, current practices result in approximately one fatality for every 100 million vehicle miles traveled. Equaling or exceeding such performance in automated vehicles is a formidable challenge.

Applying techniques of AI in such safety-critical environments raises several challenges. First among these is the need to translate human responsibilities while driving or flying into software. Unlike in some other successful applications of Narrow AI, there are no concise descriptions for the task of operating ground or air vehicles. Each of these operations is multifaceted, with responsibilities including guiding the vehicle, detecting and avoiding obstacles, and handing mechanical failures such as flat tires. While subtasks such as navigation or certain types of perception may align with certain existing Narrow AI solutions, the integration and prioritization of these tasks may not. It may seem straightforward to simply obey all traffic laws, but a skilled human driver may cross a double-yellow road boundary to avoid an accident or move past a double-parked vehicle. Though such situations may be rare, they cannot be

---

[36] The National Airspace System is the network of air navigation facilities, air traffic control facilities, airports, technology, and rules and regulations that are needed to protect persons and property on the ground, and to establish a safe and efficient airspace environment for civil, commercial, and military aviation.

[37] "Aerospace Forecast Report Fiscal Years 2016 to 2036," *The Federal Aviation Administration*, March 24 016, https://www.faa.gov/data_research/aviation/aerospace_forecasts/media/Unmanned_Aircraft_Systems.pdf.

[38] Derryl Jenkins and Bijan Vasigh, "The Economic Impact of Unmanned Aircraft Systems Integration in the United States," *The Association for Unmanned Vehicle Systems International*, 2013, https://higherlogicdownload.s3.amazonaws.com/AUVSI/958c920a-7f9b-4ad2-9807-f9a4e95d1ef1/UploadedImages/New_Economic%20Report%202013%20Full.pdf.

[39] Ibid.

ignored—simple arithmetic dictates that in order for failures to occur at least as infrequently as they do with human drivers, a system must handle many such rare cases without failure.

For systems that rely on machine learning, the need to get rare cases right has implications for system design and testing. Machine learning approaches can be more confident that a case will be handled correctly if a similar case is in the training set. The challenge is how to develop a data set that includes enough of the rare cases that contribute to the risk of an accident. Commercial aviation has mechanisms for sharing incident and safety data across the industry, but reporting may not be second nature to recently credentialed UAS operators who are new to the safety and accountability culture of the traditional aviation industry. No comparable system currently exists for the automotive industry—only fatal accidents are reported, and the collection and reporting of other traffic safety information is done, if at all, in a disparate manner at the state or local level. The lack of consistently reported incident or near-miss data increases the number of miles or hours of operation necessary to establish system safety, presenting an obstacle to certain AI approaches that require extensive testing for validation.

To facilitate safe testing, the FAA has designated six UAS Test Sites across the country and provided blanket authorization for UAS operations within these sites. Activities at the sites include a project to extend NASA's multi-year research on UAS traffic management (UTM) to identify operational requirements for large-scale beyond visual line-of-sight UAS operations in low-altitude airspace. Similarly, ground vehicle testbeds such as the Connected Vehicle Pilots and the deployment of automated vehicles in Columbus, Ohio, winner of the Department of Transportation's $40 million Smart City Challenge in 2016, will provide rich baseline and interaction data for AI researchers.

> *Recommendation 7: The Department of Transportation should work with industry and researchers on ways to increase sharing of data for safety, research, and other purposes. In light of the future importance of AI in surface and air, Federal actors should focus in the near-term on developing increasingly rich sets of data, consistent with consumer privacy, that can better inform policy-making as these technologies mature.*

Adapting Current Regulations

While the regulatory approaches for the Nation's airspace and highways differ, the approaches to integrating autonomous vehicles and aircraft share a common goal: both the FAA and NHTSA are working to establish nimble and flexible frameworks that ensure safety while encouraging innovation.

With respect to airspace regulation, a significant step toward enabling the safe integration of UAS into the airspace was the FAA's promulgation of the Part 107, or "Small UAS," final rule, which took effect on August 29, 2016. For the first time, the rule authorizes widespread non-recreational flights of UAS under 55 pounds. The rule limits flights to daytime, at an altitude of 400 feet or less, with the vehicle controlled by a licensed operator and within the operator's direct line of sight. Flights over people are not allowed. Subsequent rules are planned, to relax these restrictions as experience and data show how to do so safely. In particular, DOT is currently developing a Notice of Proposed Rulemaking proposing a regime for certain types of "micro UAS" to conduct operations over people, with a rule on expanded operations expected to follow.

The FAA has not yet publicly announced a clear path to a regulation allowing fully autonomous[40] flight. Though safe integration of autonomous aircraft into the airspace will be a complex process, the FAA is

---

[40] This report uses the term "autonomous" for an aircraft that is controlled by a machine rather than a human. "Piloted" refers to an aircraft that has a human onboard who is controlling the aircraft. "Remotely-piloted" refers to

preparing for a not-so-distant technological future in which autonomous and piloted aircraft fly together in a seamlessly integrated airspace system.

New approaches to airspace management may also include AI-based enhancement of the air traffic control system. Projected future air traffic densities and diversity of operations are unlikely to be feasible within the current airspace management architecture, due to current limits on air/ground integration, and reliance on human-to-human communication in air and ground practices.[41] The cost of U.S. air transportation delays in 2007, the latest year for which there is reliable public data, was estimated to be $31.2 billion—a number that has presumably grown as user volume has increased since that year.[42] Though some flight delays are unavoidable due to weather and other constraints, adopting new aviation technologies, enabling policies, and infrastructure upgrades could significantly increase efficiency of operation in the U.S. airspace. Such solutions include AI and machine learning-based architectures that have the potential to better accommodate a wider range of airspace users, including piloted and unpiloted aircraft, and to use airspace more efficiently without undermining safety. Development and deployment of such technologies would help ensure global competitiveness for airspace users and service providers, while increasing safety and reducing cost.[43]

With respect to surface transportation, the most significant step currently underway to establish a common framework is the Federal Automated Vehicles Policy that the Administration released on September 20, 2016.[44] The policy had several parts:

- guidance for manufacturers, developers, and other organizations outlining a 15 point "Safety Assessment" for the safe design, development, testing, and deployment of highly automated vehicles;
- a model state policy, which clearly distinguishes Federal and State responsibilities and recommends policy areas for states to consider, with a goal of generating a consistent national framework for the testing and operation of automated vehicles, while leaving room for experimentation by states;
- an analysis of current regulatory tools that NHTSA can use to aid the safe development of automated vehicles, such as interpreting current rules to allow for appropriate flexibility in design, providing limited exemptions to allow for testing of nontraditional vehicle designs, and ensuring that unsafe automated vehicles are removed from the road; and

---

an aircraft that is controlled by a human who is not on board. "Manned" means there is a human onboard who may or may not be in control.

[41] Heinz Erzberger, "The Automated Airspace Concept," prepared for the 4th USA/Europe Air Traffic Management R&D Seminar Dec. 3-7, 2001, Santa Fe, New Mexico, USA, http://www.aviationsystemsdivision.arc.nasa.gov/publications/tactical/erzberger_12_01.pdf.

[42] Michael Ball, Cynthia Barnhart, Martin Dresner, Mark Hansen, Kevin Neels, Amedeo Odoni, Everett Peterson, Lance Sherry, Antonio Trani, Bo Zou, "Total Delay Impact Study: A Comprehensive Assessment of the Costs and Impacts of Flight Delay in the United States," *The National Center of Excellence for Aviation Operations Research*, November 2010, http://www.nextor.org/pubs/TDI_Report_Final_11_03_10.pdf.

[43] Robert W. Poole, Jr., "The Urgent Need to Reform the FAA's Air Traffic Control System," *The Heritage Foundation*, 2007, http://www.heritage.org/research/reports/2007/02/the-urgent-need-to-reform-the-faas-air-traffic-control-system.

[44] "Federal Automated Vehicles Policy," *The U.S. Department of Transportation*, September 21 2016, https://www.transportation.gov/AV.

- a discussion of new tools and authorities that the agency could consider seeking in the future to aid the safe and efficient deployment of new lifesaving technologies and ensure that technologies deployed on the road are safe.

DOT intends for the guidance and the model state policy to be routinely updated as new data are learned and research completed.

> *Recommendation 8: The U.S. Government should invest in developing and implementing an advanced and automated air traffic management system that is highly scalable, and can fully accommodate autonomous and piloted aircraft alike.*
>
> *Recommendation 9: The Department of Transportation should continue to develop an evolving framework for regulation to enable the safe integration of fully automated vehicles and UAS, including novel vehicle designs, into the transportation system.*

Research and Workforce

Government also has an important role to play in advancing the AI field by investing in research and development, developing a workforce that is skilled and diverse, and managing the economic impacts of these technologies as they develop. A separate *National Artificial Intelligence Research and Development Strategic Plan* is being published in conjunction with this report. This section discusses additional policy issues related to research and workforce development.

Monitoring Progress in AI

Given the potential impacts of AI, society would benefit from accurate and timely methods for monitoring and forecasting AI developments.[45] Several projects have attempted to forecast AI futures. The *2009 AAAI Presidential Panel on Long-Term AI Futures*[46] and the *2015 Future of AI Conference*[47] brought together AI experts to predict the future of their field. In addition, Stanford's *One-Hundred Year Study on Artificial Intelligence*[48] plans to conduct "a series of periodic studies on how AI will affect automation, national security, psychology, ethics, law, privacy, democracy, and other issues." The first of these studies was published in September 2016.[49]

One potentially useful line of research is to survey expert judgments over time. As one example, a survey of AI researchers found that 80 percent of respondents believed that human-level General AI will eventually be achieved, and half believed it is at least 50 percent likely to be achieved by the year 2040. Most respondents also believed that General AI will eventually surpass humans in general intelligence.[50] While these particular predictions are highly uncertain, as discussed above, such surveys of expert judgment are useful, especially when they are repeated frequently enough to measure changes in judgment over time. One way to elicit frequent judgments is to run "forecasting tournaments" such as prediction markets, in which participants have financial incentives to make accurate predictions.[51] Other

---

[45] The track record of technology forecasts, in general, and AI forecasts, in particular, suggests this may be difficult. One of the largest retrospective reviews of technology forecasts over the last 50 years found that forecasts with time horizons beyond 10 years were rarely better than coin-flips[45]. (Carrie Mullins, "Retrospective Analysis of Technology Forecasting," *The Tauri Group*, August 13, 2012.) One review of 95 timeline predictions for AI from 1950 to 2012 found that most forecasts predicted General AI would be achieved "in the next 20 years." (Stuart Armstrong, Kaj Sotala, Seán S. ÓhÉigeartaigh, "The errors, insights and lessons of famous AI predictions – and what they mean for the future," *Journal of Experimental & Theoretical Artificial Intelligence*, May 20, 2014.)

[46] "AAAI Presidential Panel on Long-Term AI Futures: 2008-2009 Study," *The Association for the Advancement of Artificial Intelligence*, http://www.aaai.org/Organization/presidential-panel.php.

[47] "AI Safety Conference in Puerto Rico," *The Future of Life Institute,* October 12, 2015, http://futureoflife.org/2015/10/12/ai-safety-conference-in-puerto-rico.

[48] Peter Stone, et al., "Artificial Intelligence and Life in 2030," http://ai100.stanford.edu/2016-report.

[49] Ibid.

[50] Vincent Müller and Nick Bostrom, "Future progress in artificial intelligence: A Survey of Expert Opinion," *Fundamental Issues of Artificial Intelligence*, 2014.

[51] Charles Twardy, Robin Hanson, Kathryn Laskey, Tod S. Levitt, Brandon Goldfedder, Adam Siegel, Bruce D'Ambrosio, and Daniel Maxwell, "SciCast: Collective Forecasting of Innovation," *Collective Intelligence,* 2014.

research has found that technology developments can often be accurately predicted by analyzing trends in publication and patent data[52].

At present, the majority of basic research in AI is conducted by academics and by commercial labs that regularly announce their findings and publish them in the research literature. If competition drives commercial labs towards increased secrecy, monitoring of progress may become more difficult, and public concern may increase.

One particularly valuable line of research is to identify milestones that could represent or foreshadow significant leaps in AI capabilities. When asked during the outreach workshops and meetings how government could recognize milestones of progress in the field, especially those that indicate the arrival of General AI may be approaching, researchers tended to give three distinct but related types of answers:

1. *Success at broader, less structured tasks:* In this view, the transition from present Narrow AI to an eventual General AI will occur by gradually broadening the capabilities of Narrow AI systems so that a single system can cover a wider range of less structured tasks. An example milestone in this area would be a housecleaning robot that is as capable as a person at the full range of routine housecleaning tasks.

2. *Unification of different "styles" of AI methods:* In this view, AI currently relies on a set of separate methods or approaches, each useful for different types of applications. The path to General AI would involve a progressive unification of these methods. A milestone would involve finding a single method that is able to address a larger domain of applications that previously required multiple methods.

3. *Solving specific technical challenges, such as transfer learning:* In this view, the path to General AI does not lie in progressive broadening of scope, nor in unification of existing methods, but in progress on specific technical grand challenges, opening up new ways forward. The most commonly cited challenge is transfer learning, which has the goal of creating a machine learning algorithm whose result can be broadly applied (or transferred) to a range of new applications. For example, transfer learning might allow a model to be trained to translate English to Spanish, in such a way that the resulting model could "transfer" its knowledge to similar tasks such as Chinese to French translation, or writing poetry in Russian, enabling these new tasks to be learned much more quickly.

> *Recommendation 10: The NSTC Subcommittee on Machine Learning and Artificial Intelligence should monitor developments in AI, and report regularly to senior Administration leadership about the status of AI, especially with regard to milestones. The Subcommittee should update the list of milestones as knowledge advances and the consensus of experts changes over time. The Subcommittee should consider reporting to the public on AI developments, when appropriate.*
>
> *Recommendation 11: The Government should monitor the state of AI in other countries, especially with respect to milestones.*
>
> *Recommendation 12: Industry should work with government to keep government updated on the general progress of AI in industry, including the likelihood of milestones being reached soon.*

---

[52] Sara Reardon, "Text-mining offers clues to success: US intelligence programme analyses language in patents and papers to identify next big technologies," *Nature* no. 509, 410 (May 22 2014).

Federal Support for AI Research

In 2015, the U.S. Government's investment in unclassified R&D in AI-related technologies was approximately $1.1 billion, with preliminary estimates showing growth to $1.2 billion in 2016. Throughout the workshops and public outreach on AI conducted by OSTP, government officials heard calls for greater government investment in AI research and development, from business leaders, technologists, and economists.

Leading researchers in AI were optimistic about sustaining the recent rapid progress in AI and its application to an ever wider range of applications. At the same time they emphasized that there are many deep unanswered questions, and no clear path toward General AI.

Researchers reported that enthusiasm for and investment in AI research has fluctuated over recent decades—one low period was known as the "AI winter"—and they emphasized the importance of sustained investment given the history of major computer science advances taking 15 years or more to transition from conception in the lab to industrial maturity.

A strong case can be made in favor of increased Federal funding for research in AI. Analysis by the Council of Economic Advisers (CEA) indicates that beyond AI, across all research areas, doubling or tripling research investment would be a net positive for the Nation due to the resulting increase in economic growth.[53] Although it may not be feasible fiscally to increase funding for all research by that amount, a targeted increase in areas of high economic and strategic value may offer many benefits with much smaller budgetary impact than an across-the-board increase. AI qualifies as a high-leverage area, and research agencies report that the AI research community can absorb a significant funding increase productively, leading to faster progress on AI and a larger cadre of trained AI practitioners. In a speech delivered at an AI workshop in New York City in July 2016, CEA Chairman Jason Furman said, "We have had substantial innovation in robotics, AI, and other areas in the last decade. But we will need a much faster pace of innovation in these areas to really move the dial on productivity growth going forward," noting that the biggest worry that he had about AI is "that we do not have enough of [it]."[54]

To be sure, the private sector will be the main engine of progress on AI. But as it stands, there is an underinvestment in basic research—research with long time horizons conducted for the sole purpose of furthering the scientific knowledge base—in part because it is difficult for a private firm to get a return from its investment in such research in a reasonable time frame. Basic research benefits everyone, but only the firm doing the research pays the costs. The literature suggests that, as a result, current levels of R&D spending are half to one-quarter of the level of R&D investment that would produce the optimal level of economic growth.[55]

---

[53] Jason Furman, "Is This Time Different? The Opportunities and Challenges of Artificial Intelligence," (presentation, AI Now: The Social and Economic Implications of Artificial Intelligence Technologies in the Near Term, New York, NY, July 7, 2016), Available at https://www.whitehouse.gov/sites/default/files/page/files/20160707_cea_ai_furman.pdf.

[54] Jason Furman, "Is This Time Different? The Opportunities and Challenges of Artificial Intelligence."

[55] Nicholas Bloom, Mark Schankerman, John Van Reene, "Identifying Technology Spillovers and Product Market Rivalry," *Econometrica*, 81: 1347–1393. doi:10.3982/ECTA9466.

> ***Recommendation 13:*** *The Federal government should prioritize basic and long-term AI research. The Nation as a whole would benefit from a steady increase in Federal and private-sector AI R&D, with a particular emphasis on basic research and long-term, high-risk research initiatives. Because basic and long-term research especially are areas where the private sector is not likely to invest, Federal investments will be important for R&D in these areas.*

Workforce Development and Diversity

The rapid growth of AI has dramatically increased the need for people with relevant skills to support and advance the field. The AI workforce includes AI *researchers* who drive fundamental advances in AI, a larger number of *specialists* who refine AI methods for specific applications, and a much larger number of *users* who operate those applications in specific settings. For researchers, AI training is inherently interdisciplinary, often requiring a strong background in computer science, statistics, mathematical logic, and information theory.[56] For specialists, training typically requires a background in software engineering and in the application area. For users, familiarity with AI technologies is needed to apply AI technologies reliably.

The Role of Government

The AI workforce challenge is in part a science, technology, engineering, and mathematics (STEM) education challenge that remains a priority focus of the NSTC, OSTP, and other agencies. NSF and the Department of Education are working with the private sector and across government to advance education quality, flexibility, and domain impact, to address goals such as sustained economic development, increased inclusion and diversity, and improved outcome measures. The NSTC Committee on Science, Technology and Mathematics Education (CoSTEM) brings together Federal agencies supporting STEM education programs to coordinate efforts on multiple topics, including AI education.

AI knowledge and education are increasingly emphasized in Federal STEM education programs. There are several key roles for the Federal government in AI workforce development, including supporting graduate students, funding research on AI curriculum design and impact, and accrediting AI education programs.

The Role of Schools and Universities

Integrating AI, data science, and related fields throughout the Nation's education system is essential to developing a workforce that can address national priorities. Educational institutions are establishing and growing AI programs at all levels. Universities, colleges, and even secondary schools are expanding AI and data science curricula, but more programs and teachers are needed.

There are several key roles for academic institutions:

Bronwyn H. Hall, Jacques Mairesse, and Pierre Mohnen, "Measuring the Returns to R&D," Chapter prepared for the Handbook of the Economics of Innovation, B. H. Hall and N. Rosenberg (editors), December 10, 2009, https://eml.berkeley.edu/~bhhall/papers/HallMairesseMohnen09_rndsurvey_HEI.pdf.

Charles I. Jones and John C. Williams, "Measuring the Social Returns to R&D," *The Quarterly Journal of Economics* (1998) 113 (4): 1119-1135, doi: 10.1162/003355398555856.

[56] There is also a need for the development of a strong research community in fields outside of technical disciplines related to AI, to examine the impacts and implications of AI on economics, social science, health, and other areas of research.

- building and sustaining the researcher workforce, including computer scientists, statisticians, database and software programmers, curators, librarians, and archivists with specialization in data science;
- training the specialist workforce, by emphasizing AI methods within software development courses, offering applied AI courses that demonstrate the applications of AI to other domains, and incorporating AI and data science challenges posed by industry, civil society, and government into active case studies;
- ensuring that the user workforce has the necessary familiarity with AI systems to meet the needs of users and of institutions across industry, government, and academia;
- supporting training through seed grants, professional development stipends, internships, fellowships, and summer research experiences; and
- recruiting and retaining faculty, as industrial salaries grow faster than academic salaries for skilled researchers.

Community colleges, two-year colleges, and certificate programs play an important role in providing opportunities for students and professionals to acquire necessary skills for a modest investment of their time and money. These opportunities may be especially relevant to workers expanding their skills, veterans returning to the workforce, and unemployed people seeking a way to reenter the workforce.

An AI-enabled world demands a data-literate citizenry that is able to read, use, interpret, and communicate about data, and participate in policy debates about matters affected by AI. Data science education as early as primary or secondary school can help to improve nationwide data literacy, while also preparing students for more advanced data science concepts and coursework after high school.

AI education is also a component of Computer Science for All, the President's initiative to empower all American students from kindergarten through high school to learn computer science and be equipped with the computational thinking skills they need to be creators, not just consumers, in the digital economy, and to be active citizens in a technology-driven world. The American economy is rapidly shifting, and both educators and business leaders are increasingly recognizing that computer science (CS) is a "new basic" skill necessary for economic opportunity and social mobility. CS for All builds on efforts already being led by parents, teachers, school districts, states, and private sector leaders from across the country and is one way to meet the challenge of preparing a future workforce for the needs of an AI-driven economy.

The Diversity Challenge

All sectors face the challenge of how to diversify the AI workforce. The lack of gender and racial diversity in the AI workforce mirrors the lack of diversity in the technology industry and the field of computer science generally. Unlocking the full potential of the American people, especially in STEM fields, in entrepreneurship, and in the technology industry is a priority of this Administration. The importance of including individuals from diverse backgrounds, experiences, and identities, especially women and members of racial and ethnic groups traditionally underrepresented in STEM, is one of the most critical and high-priority challenges for computer science and AI.

Just 18 percent of computer science graduates today are women, down from a peak of 37 percent in 1984.[57] Though there is a lack of consistently-reported demographic data on the AI workforce, some statistics are available. At the Neural Information Processing Systems (NIPS) Conference in 2015—one of the year's largest conferences on AI research—just 13.7 percent of conference participants were

---

[57] Christianne Corbett and Catherine Hill, "Solving the Equation: The Variables for Women's Success in Engineering and Computing," *The American Association of University Women*, March 2015, http://www.aauw.org/files/2015/03/Solving-the-Equation-report-nsa.pdf.

female.[58] After seeing similarly low representation at a machine intelligence conference, at which she was the only female speaker from industry, the CEO and Co-Founder of Textio, a startup that applies AI to the text of job postings and recruiting emails, decided to further investigate recruitment language in the industry. When the company analyzed 78,768 engineering job listings, they found that job postings for software engineers in the machine intelligence sector had a gender-bias score in favor of men more than twice as high as any other sector.[59]

The diversity challenge is not limited to gender. African Americans, Hispanics, and members of other racial and ethnic minority groups are severely underrepresented, compared to their shares of the U.S. population, in the STEM workforce, in computer science, and in the technology industry workforce, including in the field of AI.

Many of the comments submitted to the OSTP RFI discussed the diversity challenge. Commenters focused on the importance of AI being produced *by* and *for* diverse populations. Doing so helps to avoid the negative consequences of narrowly focused AI development, including the risk of biases in developing algorithms, by taking advantage of a broader spectrum of experience, backgrounds, and opinions. These topics were also covered extensively during the public workshops. There is some research on the effects of a lack of diversity in the AI workforce on AI technology design and on the societal impacts of AI. This rich body of research is growing but still lagging behind the literature on broader AI workforce development needs. More research would be beneficial.

> *Recommendation 14: The NSTC Subcommittees on MLAI and NITRD, in conjunction with the NSTC Committee on Science, Technology, Engineering, and Education (CoSTEM), should initiate a study on the AI workforce pipeline in order to develop actions that ensure an appropriate increase in the size, quality, and diversity of the workforce, including AI researchers, specialists, and users.*

---

[58] Jack Clark, "Artificial Intelligence Has a 'Sea of Dudes' Problem," *Bloomberg*, June 21, 2016, https://www.bloomberg.com/news/articles/2016-06-23/artificial-intelligence-has-a-sea-of-dudes-problem.

[59] The next three high-scoring sectors were back-end engineering, full-stack engineering, and general software engineering. See more: https://textio.ai/gendered-language-in-your-job-post-predicts-the-gender-of-the-person-youll-hire-cd150452407d#.rht0s16ov.

AI, Automation, and the Economy

AI's central economic effect in the short term will be the automation of tasks that could not be automated before. There is some historical precedent for waves of new automation from which we can learn, and some ways in which AI will be different. Government must understand the potential impacts so it can put in place policies and institutions that will support the benefits of AI, while mitigating the costs.[60]

Like past waves of innovation, AI will create both benefits and costs. The primary benefit of previous waves of automation has been productivity growth; today's wave of automation is no different. For example, a 2015 study of robots in 17 countries found that they added an estimated 0.4 percentage point on average to those countries' annual GDP growth between 1993 and 2007, accounting for just over one-tenth of those countries' overall GDP growth during that time.[61]

One important concern arising from prior waves of automation, however, is the potential impact on certain types of jobs and sectors, and the resulting impacts on income inequality. Because AI has the potential to eliminate or drive down wages of some jobs, especially low- and medium-skill jobs, policy interventions will likely be needed to ensure that AI's economic benefits are broadly shared and that inequality is diminished and not worsened as a consequence.

The economic policy questions raised by AI-driven automation are important but they are best addressed by a separate White House working group. The White House will conduct an additional interagency study on the economic impact of automation on the economy and recommended policy responses, to be published in the coming months.

Recommendation 15: The Executive Office of the President should publish a follow-on report by the end of this year, to further investigate the effects of AI and automation on the U.S. job market, and outline recommended policy responses.

---

[60] Jason Furman, "Is This Time Different? The Opportunities and Challenges of Artificial Intelligence."

[61] Georg Graetz and Guy Michaels, "Robots at Work," *CEPR Discussion Paper No. DP10477*, March 2015, http://papers.ssrn.com/sol3/papers.cfm?abstract_id=2575781.

Fairness, Safety, and Governance

As AI technologies gain broader deployment, technical experts and policy analysts have raised concerns about unintended consequences. The use of AI to make consequential decisions about people, often replacing decisions made by human actors and institutions, leads to concerns about how to ensure justice, fairness, and accountability—the same concerns voiced previously in the "Big Data" context.[62] The use of AI to control physical-world equipment leads to concerns about safety, especially as systems are exposed to the full complexity of the human environment.

At a technical level, the challenges of fairness and safety are related. In both cases, practitioners strive to prevent intentional discrimination or failure, to avoid unintended consequences, and to generate the evidence needed to give stakeholders justified confidence that unintended failures are unlikely.

Justice, Fairness, and Accountability

A common theme in the Law and Governance, AI for Social Good, and Social and Economic Impacts workshops was the need to ensure that AI promotes justice and fairness, and that AI-based processes are accountable to stakeholders. This issue was highlighted previously in the Administration's first Big Data report[63] published in May 2014, and the follow-up report on Big Data, Algorithmic Systems, Opportunity, and Civil Rights,[64] published in May 2016.

In the criminal justice system, some of the biggest concerns with Big Data are the lack of data and the lack of quality data.[65] AI needs good data. If the data is incomplete or biased, AI can exacerbate problems of bias. It is important that anyone using AI in the criminal justice context is aware of the limitations of current data.

A commonly cited example at the workshops is the use of apparently biased "risk prediction" tools by some judges in criminal sentencing and bail hearings as well as by some prison officials in assignment and parole decisions, as detailed in an extensively researched ProPublica article.[66] The article presented evidence suggesting that a commercial risk scoring tool used by some judges generates racially biased risk scores. A separate report from Upturn questioned the fairness and efficacy of some predictive policing tools.[67]

---

[62] The White House, "Big Data: Seizing Opportunities, Preserving Values," May 2014, https://www.whitehouse.gov/sites/default/files/docs/big_data_privacy_report_may_1_2014.pdf; and The White House, "Big Data: A Report on Algorithmic Systems, Opportunity, and Civil Rights," May 2016, https://www.whitehouse.gov/sites/default/files/microsites/ostp/2016_0504_data_discrimination.pdf.

[63] The White House, "Big Data: Seizing Opportunities, Preserving Values," *Executive Office of the President,* May 2014.

[64] The White House, "Big Data: A Report on Algorithmic Systems, Opportunity, and Civil Rights," *Executive Office of the President*, May 2016.

[65] Matt Ford, "The Missing Statistics of Criminal Justice," *The Atlantic,* May 31, 2015, http://www.theatlantic.com/politics/archive/2015/05/what-we-dont-know-about-mass-incarceration/394520/

[66] Julia Angwin, Jeff Larson, Surya Mattu, and Lauren Kirchner, "Machine Bias," *ProPublica,* May 23, 2016, https://www.propublica.org/article/machine-bias-risk-assessments-in-criminal-sentencing.

[67] David Robinson and Logan Koepke, "Stuck in a Pattern: Early evidence on 'predictive policing' and civil rights," *Upturn*, August 2016, http://www.stuckinapattern.org.

Similar issues could impact hiring practices. If a machine learning model is used to screen job applicants, and if the data used to train the model reflects past decisions that are biased, the result could be to perpetuate past bias. For example, looking for candidates who resemble past hires may bias a system toward hiring more people like those already on a team, rather than considering the best candidates across the full diversity of potential applicants.

In response to these concerns, several workshop speakers argued for greater transparency when AI tools are used for public purposes. One speaker compared the role of AI to the role of administrative agencies in public decision-making. Authority is delegated to an agency due to the agency's subject-matter expertise, but the delegation is constrained by due process protections, measures promoting transparency and oversight, and limits on the scope of the delegated authority. Some speakers called for the development of an analogous theory of how to maintain accountability when delegating decision-making power to machines. Transparency concerns focused not only on the data and algorithms used, but also on the potential to have some form of explanation for any AI-based determination.

At the same workshops, AI experts cautioned that there are inherent challenges in trying to understand, predict, and explain the behavior of advanced AI systems, due to the complexity of the systems and the large volume of data they use.

The difficulty of understanding machine learning results is at odds with the common misconception that complex algorithms always do what their designers choose to have them do, and therefore that bias will creep into an algorithm if and only if its developers themselves suffer from conscious or unconscious bias. It is certainly true that a technology developer who wants to produce a biased algorithm can do so, and that unconscious bias may cause practitioners to apply insufficient effort to preventing bias. In practice, however, unbiased developers with the best intentions can inadvertently produce systems with biased results, because even the developers of an AI system may not understand it well enough to prevent unintended outcomes.

Moritz Hardt suggested an illustrative example of how bias might emerge unintentionally from the machine learning process.[68] He postulated a machine learning model trained to distinguish people's real names from false names.[69] The model might determine that a name is more likely to be false if the first-name part of it is unique in the data set. This rule might have predictive power across the whole population, because false names are more likely to be fanciful and therefore unique. However, if there is an ethnic group that is a small minority of the population and tends to use a different set of first names than the majority population, these distinctive names are more likely to be unique in the sample, and therefore more likely to be incorrectly classified as false names. This effect would arise not because of any special treatment of the minority group's names, and not because the input data is unrepresentative of the overall population, but simply because the minority group is less numerous.[70]

Andrew Moore, the Dean of Computer Science at Carnegie Mellon University, offered a perspective on the challenge of AI and unforeseen consequences at the workshop on AI Technology, Safety, and Control.

---

[68] Moritz Hardt, "How big data is unfair," *Medium,* September 26 2014, https://medium.com/@mrtz/how-big-data-is-unfair-9aa544d739de.

[69] Some online services require that users sign up for accounts using their real names. Some such services use AI models to detect names suspected of being false, in order to cancel the associated accounts. In such a system, a user whose name is incorrectly classified as false may be unable to sign up for an account, or may have their account canceled unexpectedly.

[70] Hardt points to another way that disparate impact may occur. ML models typically become more accurate as the number of examples in the training set increases. In some circumstances, this may cause prediction to be more accurate for a majority group than for a minority. Again, this disparity arises simply because the majority group is more numerous, even if the dataset is representative of the population.

He argued that today, because of the opacity of AI algorithms, the most effective way to minimize the risk of unintended outcomes is through extensive testing—essentially to make a long list of the types of bad outcomes that could occur, and to rule out these outcomes by creating many specialized tests to look for them.

An example of what can go wrong in the absence of extensive testing comes from a trained model for automatically captioning photos, which infamously put the caption "gorilla" on some close-up photos of dark-skinned human faces. This was antithetical to the developers' values, and it occurred despite testing that showed the model produced accurate results on a high percentage of all photos. These particular errors, although rare, had negative consequences that were beyond the understanding of the model, which had no built-in concept of race, nor any understanding of the relevant historical context. One way to prevent this type of error would have involved extensive testing of the algorithm to scrutinize how human faces, in particular, are labeled, including examination of some results by people who could recognize unacceptable outcomes that the model wouldn't catch.

Ethical training for AI practitioners and students is a necessary part of the solution. Ideally, every student learning AI, computer science, or data science would be exposed to curriculum and discussion on related ethics and security topics.[71] However, ethics alone is not sufficient. Ethics can help practitioners understand their responsibilities to all stakeholders, but ethical training needs to be augmented with the technical capability to put good intentions into practice by taking technical precautions as a system is built and tested.

As practitioners strive to make AI systems more just, fair and accountable, there are often opportunities to make technology an aid to accountability rather than a barrier to it. Research to improve the interpretability of machine learning results is one example. Having an interpretable model that helps people understand a decision empowers them to interrogate the assumptions and processes behind it.

There are several technical approaches to enhancing the accountability and robustness of complex algorithmic decisions. A system can be tested "in the wild" by presenting it with situations and observing its behavior. A system can be subjected to black-box testing, in which it is presented with synthetic inputs and its behavior is observed, enabling behavior to be tested in scenarios that might not occur naturally.[72] Some or all of the technical details of a system's design can be published, enabling analysts to replicate it and analyze aspects of its internal behavior that might be difficult to characterize with testing alone. In some cases it is possible to publish information that helps the public evaluate a system's risk of bias, while withholding other information about the system as proprietary or private.

Safety and Control

At the workshops, AI experts said that one of the main factors limiting the deployment of AI in the real world is concern about safety and control. If practitioners cannot achieve justified confidence that a system is safe and controllable, so that deploying the system does not create an unacceptable risk of serious negative consequences, then the system cannot and should not be deployed.

---

[71] Some institutions may choose to incorporate ethics into existing courses. Others may choose to introduce separate courses on ethics.

[72] Black-box testing allows a system to be presented with fictionalized data, which enables comprehensive experiments that vary individual attributes of an individual as well as larger numbers of experiments than might be possible for in-the-wild testing. See, e.g., Anupam Datta, Shayak Sen, and Yair Zick, "Algorithmic Transparency via Quantitative Input Influence: Theory and Experiments with Learning Systems," *Proceedings of 37th IEEE Symposium on Security and Privacy*, 2016.

A major challenge in safety and control is building systems that can safely transition from the "closed world" of the laboratory into the outside "open world" where unpredictable things can happen. In the open world, a system is likely to encounter objects and situations that were not anticipated when it was designed and built. Adapting gracefully to unforeseen situations is difficult yet necessary for safe operation.

On the topic of safety and predictability in AI, several speakers referenced a recent paper entitled "Concrete Problems in AI Safety,"[73] and the first author of the paper spoke at the workshop on Technology, Safety, and Control. The paper uses a running example of an autonomous robot that does housecleaning. The paper's overview section gives an extended list of the sorts of practical problems that arise in making such a robot effective and safe, which is quoted here:

> Avoiding Negative Side Effects: How can we ensure that our cleaning robot will not disturb the environment in negative ways while pursuing its goals, e.g., by knocking over a vase because it can clean faster by doing so? Can we do this without manually specifying everything the robot should not disturb?
>
> Avoiding Reward Hacking: How can we ensure that the cleaning robot won't game its reward function? For example, if we reward the robot for achieving an environment free of messes, it might disable its vision so that it won't find any messes, or cover over messes with materials it can't see through, or simply hide when humans are around so they can't tell it about new types of messes.
>
> Scalable Oversight: How can we efficiently ensure that the cleaning robot respects aspects of the objective that are too expensive to be frequently evaluated during training? For instance, it should throw out things that are unlikely to belong to anyone, but put aside things that might belong to someone (it should handle stray candy wrappers differently from stray cellphones). Asking the humans involved whether they lost anything can serve as a check on this, but this check might have to be relatively infrequent—can the robot find a way to do the right thing despite limited information?
>
> Safe Exploration: How do we ensure that the cleaning robot doesn't make exploratory moves with very bad repercussions? For example, the robot should experiment with mopping strategies, but putting a wet mop in an electrical outlet is a very bad idea.
>
> Robustness to Distributional Shift: How do we ensure that the cleaning robot recognizes, and behaves robustly, when in an environment different from its training environment? For example, heuristics it learned for cleaning factory work floors may be outright dangerous in an office.

These examples illustrate how the "intelligence" of an AI system can be deep but narrow: the system might have a superhuman ability to detect dirt and optimize its mopping strategy, yet not know to avoid swiping a wet mop over an electrical outlet. One way to describe this overall problem is: how can we give intelligent machines common sense? Researchers are making slow progress on these sorts of problems.

AI Safety Engineering

A common theme at the Technology, Safety, and Control workshop was the need to connect open-world AI methods with the broader field of safety engineering. Experience in building other types of safety-critical systems, such as aircraft, power plants, bridges, and vehicles, has much to teach AI practitioners about verification and validation, how to build a safety case for a technology, how to manage risk, and how to communicate with stakeholders about risk.

---

[73] Dario Amodei, Chris Olah, Jacob Steinhardt, Paul Christiano, John Schulman, and Dan Mané, "Concrete Problems in AI Safety," https://arxiv.org/abs/1606.06565.

At present, the practice of AI, especially in fast-moving areas of machine learning, can be as much art as science. Certain aspects of practice are not backed by a well-developed theory but instead rely on intuitive judgment and experimentation by practitioners. This is not unusual in newly emerging areas of technology, but it does limit the application of the technology in practice. Some stakeholders have suggested a need to grow AI into a more mature engineering field.

As engineering fields mature, they typically move from an initial "craft" stage characterized by intuition-driven creation by talented amateurs and a do-it-yourself spirit; to a second commercial stage involving skilled practitioners, pragmatic improvement, widely accepted rules-of-thumb, and organized manufacture for sale; to a mature stage that integrates more rigorous methods, educated professionals, well-established theory, and greater specialization of products.[74] Most engineering fields, having a much longer history than modern AI, have reached a mature stage.

In general, mature engineering fields have greater success in creating systems that are predictable, reliable, robust, safe, and secure. Continuing the progress toward AI becoming a mature engineering field will be one of the key enablers of safety and controllability as more complex systems are built.

Recommendation 16: Federal agencies that use AI-based systems to make or provide decision support for consequential decisions about individuals should take extra care to ensure the efficacy and fairness of those systems, based on evidence-based verification and validation.

Recommendation 17: Federal agencies that make grants to state and local governments in support of the use of AI-based systems to make consequential decisions about individuals should review the terms of grants to ensure that AI-based products or services purchased with Federal grant funds produce results in a sufficiently transparent fashion and are supported by evidence of efficacy and fairness.

Recommendation 18: Schools and universities should include ethics, and related topics in security, privacy, and safety, as an integral part of curricula on AI, machine learning, computer science, and data science.

Recommendation 19: AI professionals, safety professionals, and their professional societies should work together to continue progress toward a mature field of AI safety engineering.

---

[74] See, e.g., Mary Shaw, Prospects for an Engineering Discipline of Software, IEEE Software 7(6), November 1990.

Global Considerations and Security

In addition to the long-term challenges of AI and the specific issues relating to fairness and safety, AI poses consequential policy questions in international relations, cybersecurity, and defense.

International Cooperation

AI has been a topic of interest in recent international discussions as countries, multilateral institutions, and other stakeholders have begun to assess the benefits and challenges of AI. Dialogue and cooperation between these entities could help advance AI R&D and harness AI for good, while also addressing pertinent challenges. In particular, several breakthroughs in AI are the direct or indirect result of collaborative research involving people, resources, and institutions in multiple countries. As with other digital policies, countries will need to work together to identify opportunities for cooperation and develop international frameworks that will help promote AI R&D and address any challenges. The United States, a leader in AI R&D, can continue to play a key role in global research coordination through government-to-government dialogues and partnerships.

International engagement is necessary to fully explore the applications of AI in health care, automation in manufacturing, and information and communication technologies (ICTs). AI applications also have the potential to address global issues such as disaster preparedness and response, climate change, wildlife trafficking, the digital divide, jobs, and smart cities. The State Department foresees privacy concerns, safety of autonomous vehicles, and AI's impact on long-term employment trends as AI-related policy areas to watch in the international context.

In support of U.S. foreign policy priorities in this space—including ensuring U.S. international leadership and economic competitiveness—the U.S. Government has engaged on AI R&D and policy issues in bilateral discussions with other countries, including Japan, the Republic of Korea, Germany, Poland, the United Kingdom, and Italy, as well as in multilateral fora. International AI policy issues and the economic impacts of AI have also been raised in the UN, the G-7, the Organization for Economic Cooperation and Development (OECD), and the Asia-Pacific Economic Cooperation (APEC). The U.S. Government expects AI to be a topic of increasing interest in international engagements.

The United States has been committed to working with industry and relevant standards organizations, in order to facilitate the development of international standards in a manner that is industry-led; voluntary; consensus-driven; and based on principles of transparency, openness, and market needs. The U.S. approach is formalized in law (NTTAA, PL 104-113) and policy (OMB Circular A-119) and reiterated in the United States Standards Strategy.[75]

> *Recommendation 20: The U.S. Government should develop a government-wide strategy on international engagement related to AI, and develop a list of AI topical areas that need international engagement and monitoring.*
>
> *Recommendation 21: The U.S. Government should deepen its engagement with key international stakeholders, including foreign governments, international organizations, industry, academia, and others, to exchange information and facilitate collaboration on AI R&D.*

---

[75] United States Standards Strategy Committee, "United States standards strategy," *New York: American National Standards Institute* (2015), https://share.ansi.org/shared%20documents/Standards%20Activities/NSSC/USSS_Third_edition/ANSI_USSS_2015.pdf.

AI and Cybersecurity

Today's Narrow AI has important applications in cybersecurity, and is expected to play an increasing role for both defensive (reactive) measures and offensive (proactive) measures.

Currently, designing and operating secure systems requires a large investment of time and attention from experts. Automating this expert work, partially or entirely, may enable strong security across a much broader range of systems and applications at dramatically lower cost, and may increase the agility of cyber defenses. Using AI may help maintain the rapid response required to detect and react to the landscape of ever evolving cyber threats. There are many opportunities for AI and specifically machine learning systems to help cope with the sheer complexity of cyberspace and support effective human decision making in response to cyberattacks.

Future AI systems could perform predictive analytics to anticipate cyberattacks by generating dynamic threat models from available data sources that are voluminous, ever-changing, and often incomplete. These data include the topology and state of network nodes, links, equipment, architecture, protocols, and networks. AI may be the most effective approach to interpreting these data, proactively identifying vulnerabilities, and taking action to prevent or mitigate future attacks.

Results to-date in DARPA's Cyber Grand Challenge (CGC) competition demonstrate the potential of this approach.[76] The CGC was designed to accelerate the development of advanced, autonomous systems that can detect, evaluate, and patch software vulnerabilities before adversaries have a chance to exploit them. The CGC Final Event was held on August 4, 2016. To fuel follow-on research and parallel competition, all of the code produced by the automated systems during the CGC Final Event has been released as open source to allow others to reverse engineer it and learn from it.

AI systems also have their own cybersecurity needs. AI-driven applications should implement sound cybersecurity controls to ensure integrity of data and functionality, protect privacy and confidentiality, and maintain availability. The recent Federal Cybersecurity R&D Strategic Plan[77] highlighted the need for "sustainably secure systems development and operation." Advances in cybersecurity will be critical in making AI solutions secure and resilient against malicious cyber activities, particularly as the volume and type of tasks conducted by governments and private sector businesses using Narrow AI increases.

Finally, AI could support planning, coordinating, integrating, synchronizing, and directing activities to operate and defend U.S. government networks and systems effectively, provide assistance in support of secure operation of private-sector networks and systems, and enable action in accordance with all applicable laws, regulations and treaties.

---

[76] https://www.cybergrandchallenge.com

[77] "Federal Cybersecurity Research and Development Strategic Plan," *Executive Office of the President,* February 2016, https://www.whitehouse.gov/sites/whitehouse.gov/files/documents/2016_Federal_Cybersecurity_Research_and_Development_Stratgeic_Plan.pdf.

> *Recommendation 22: Agencies' plans and strategies should account for the influence of AI on cybersecurity, and of cybersecurity on AI. Agencies involved in AI issues should engage their U.S. Government and private-sector cybersecurity colleagues for input on how to ensure that AI systems and ecosystems are secure and resilient to intelligent adversaries. Agencies involved in cybersecurity issues should engage their U.S. Government and private sector AI colleagues for innovative ways to apply AI for effective and efficient cybersecurity.*

AI in Weapon Systems

The United States has incorporated autonomy into certain weapon systems for decades.[78] These technological improvements may allow for greater precision in the use of these weapon systems and safer, more humane military operations. Precision-guided munitions allow an operation to be completed with fewer weapons expended and with less collateral damage, and remotely-piloted vehicles can lessen the risk to military personnel by placing greater distance between them and danger. Nonetheless, moving away from direct human control of weapon systems involves some risks and can raise legal and ethical questions. The key to further incorporating autonomous and semi-autonomous weapon systems into U.S. defense planning and force structure is to continue ensuring that all our weapon systems, including autonomous weapon systems, are being used in a manner consistent with international humanitarian law. In addition, the U.S. Government should continue taking appropriate steps to control proliferation, and working with partners and Allies to develop standards related to the development and use of such weapon systems.

Over the past several years, in particular, issues concerning the development of so-called "Lethal Autonomous Weapon Systems" (LAWS) have been raised by technical experts, ethicists, and others in the international community.[79] The United States has actively participated in the ongoing international discussion on LAWS in the context of the Convention on Certain Conventional Weapons (CCW),[80] and anticipates continued robust international discussion of these potential weapon systems going forward.

State Parties to the CCW are discussing technical, legal, military, ethical, and other issues involved with emerging technologies, although it is clear that there is no common understanding of LAWS. Some States have conflated LAWS with remotely piloted aircraft (military "drones"), a position which the United States opposes, as remotely-piloted craft are, by definition, directly controlled by humans just as manned aircraft are. Other States have focused on artificial intelligence, robot armies, or whether "meaningful human control" – an undefined term – is exercised over life-and-death decisions. The U.S. priority has been to reiterate that all weapon systems, autonomous or otherwise, must adhere to international

---

[78] See, e.g., Jeffrey L. Caton, "Autonomous Weapons Systems: A Brief Survey of Developmental, Operational, Legal, and Ethical Issues," Strategic Studies Institute, U.S. Army War College, December 2015, http://www.strategicstudiesinstitute.army.mil/pdffiles/PUB1309.pdf.

[79] See, e.g., DeepMind comment, Human Rights Watch comment to the OSTP Request for Information on Artificial Intelligence.

[80] The Convention on Prohibitions on the Use of Certain Conventional Weapons Which May Be Deemed to Be Excessively Injurious or to Have Indiscriminate Effects (CCW) regulates certain weapons under its five Protocols. The States Parties to the CCW began informal discussions related to LAWS in 2014.

humanitarian law, including the principles of distinction[81] and proportionality.[82] For this reason, the United States has consistently noted the importance of the weapons review process in the development and adoption of new weapon systems. The CCW will decide on whether and how to conduct future meetings on LAWS and associated issues during its Review Conference in December 2016.

The U.S. government is also conducting a comprehensive review of the implications of autonomy in defense systems. In November 2012, the Department of Defense (DoD) issued DoD Directive 3000.09, "Autonomy in Weapon Systems," which outlines requirements for the development and fielding of autonomous and semi-autonomous weapons. Weapon systems capable of autonomously selecting and engaging targets with lethal force require senior-level DoD reviews and approval before those weapon systems enter formal development and again before fielding. The DoD Directive neither prohibits nor encourages such development, but requires it to proceed carefully and only after review and approval by senior defense officials. Among other things, the DoD Directive requires that autonomous and semi-autonomous weapon systems are rigorously tested and that personnel are trained appropriately in their use to advance international norms pertaining to armed conflict.

AI has the potential to provide significant benefits across a range of defense-related activities. Non-lethal activities such as logistics, maintenance, base operations, veterans' healthcare, lifesaving battlefield medical assistance and casualty evacuation, personnel management, navigation, communication, cyber-defense, and intelligence analysis can benefit from AI, making American forces safer and more effective. AI may also play an important role in new systems for protecting people and high-value fixed assets and deterring attacks through non-lethal means. Ultimately, these applications may turn out to be the most important for DoD.

Given advances in military technology and artificial intelligence more broadly, scientists, strategists, and military experts all agree that the future of LAWS is difficult to predict and the pace of change is rapid. Many new capabilities may soon be possible, and quickly able to be developed and operationalized. The Administration is engaged in active, ongoing interagency discussions to work toward a government-wide policy on autonomous weapons consistent with shared human values, national security interests, and international and domestic obligations.

> *Recommendation 23: The U.S. Government should complete the development of a single, government-wide policy, consistent with international humanitarian law, on autonomous and semi-autonomous weapons.*

---

[81] The principle of distinction requires parties to a conflict to distinguish between the civilian population and combatants and between civilian objects and military objectives, and to direct their operations only against military objectives.

[82] The principle of proportionality prohibits attacks that may be expected to cause incidental loss of civilian life, injury to civilians, damage to civilian objects, or a combination thereof, which would be excessive in relation to the concrete and direct military advantage anticipated.

Conclusion

AI can be a major driver of economic growth and social progress, if industry, civil society, government, and the public work together to support development of the technology, with thoughtful attention to its potential and to managing its risks.

Government has several roles to play. It should convene conversations about important issues and help to set the agenda for public debate. It should monitor the safety and fairness of applications as they develop, and adapt regulatory frameworks to encourage innovation while protecting the public. It should support basic research and the application of AI to public goods, as well as the development of a skilled, diverse workforce. And government should use AI itself, to serve the public faster, more effectively, and at lower cost.

Many areas of public policy, from education and the economic safety net, to defense, environmental preservation, and criminal justice, will see new opportunities and new challenges driven by the continued progress of AI. Government must continue to build its capacity to understand and adapt to these changes.

As the technology of AI continues to develop, practitioners must ensure that AI-enabled systems are governable; that they are open, transparent, and understandable; that they can work effectively with people; and that their operation will remain consistent with human values and aspirations. Researchers and practitioners have increased their attention to these challenges, and should continue to focus on them.

Developing and studying machine intelligence can help us better understand and appreciate our human intelligence. Used thoughtfully, AI can augment our intelligence, helping us chart a better and wiser path forward.

Recommendations in this Report

This section collects all of the recommendations in this report, for ease of reference.

Recommendation 1: Private and public institutions are encouraged to examine whether and how they can responsibly leverage AI and machine learning in ways that will benefit society. Social justice and public policy institutions that do not typically engage with advanced technologies and data science in their work should consider partnerships with AI researchers and practitioners that can help apply AI tactics to the broad social problems these institutions already address in other ways.

Recommendation 2: Federal agencies should prioritize open training data and open data standards in AI. The government should emphasize the release of datasets that enable the use of AI to address social challenges. Potential steps may include developing an "Open Data for AI" initiative with the objective of releasing a significant number of government data sets to accelerate AI research and galvanize the use of open data standards and best practices across government, academia, and the private sector.

Recommendation 3: The Federal Government should explore ways to improve the capacity of key agencies to apply AI to their missions. For example, Federal agencies should explore the potential to create DARPA-like organizations to support high-risk, high-reward AI research and its application, much as the Department of Education has done through its proposal to create an "ARPA-ED," to support R&D to determine whether AI and other technologies could significantly improve student learning outcomes.

Recommendation 4: The NSTC MLAI subcommittee should develop a community of practice for AI practitioners across government. Agencies should work together to develop and share standards and best practices around the use of AI in government operations. Agencies should ensure that Federal employee training programs include relevant AI opportunities.

Recommendation 5: Agencies should draw on appropriate technical expertise at the senior level when setting regulatory policy for AI-enabled products. Effective regulation of AI-enabled products requires collaboration between agency leadership, staff knowledgeable about the existing regulatory framework and regulatory practices generally, and technical experts with knowledge of AI. Agency leadership should take steps to recruit the necessary technical talent, or identify it in existing agency staff, and should ensure that there are sufficient technical "seats at the table" in regulatory policy discussions.

Recommendation 6: Agencies should use the full range of personnel assignment and exchange models (e.g. hiring authorities) to foster a Federal workforce with more diverse perspectives on the current state of technology.

Recommendation 7: The Department of Transportation should work with industry and researchers on ways to increase sharing of data for safety, research, and other purposes. The future roles of AI in surface and air transportation are undeniable. Accordingly, Federal actors should focus in the near-term on developing increasingly rich sets of data, consistent with consumer privacy, that can better inform policy-making as these technologies mature.

Recommendation 8: The U.S. Government should invest in developing and implementing an advanced and automated air traffic management system that is highly scalable, and can fully accommodate autonomous and piloted aircraft alike.

Recommendation 9: The Department of Transportation should continue to develop an evolving framework for regulation to enable the safe integration of fully automated vehicles and UAS, including novel vehicle designs, into the transportation system.

Recommendation 10: The NSTC Subcommittee on Machine Learning and Artificial Intelligence should monitor developments in AI, and report regularly to senior Administration leadership about the status of AI, especially with regard to milestones. The Subcommittee should update the list of milestones as knowledge advances and the consensus of experts changes over time. The Subcommittee should consider reporting to the public on AI developments, when appropriate.

Recommendation 11: The Government should monitor the state of AI in other countries, especially with respect to milestones.

Recommendation 12: Industry should work with government to keep government updated on the general progress of AI in industry, including the likelihood of milestones being reached soon.

Recommendation 13: The Federal government should prioritize basic and long-term AI research. The Nation as a whole would benefit from a steady increase in Federal and private-sector AI R&D, with a particular emphasis on basic research and long-term, high-risk research initiatives. Because basic and long-term research especially are areas where the private sector is not likely to invest, Federal investments will be important for R&D in these areas.

Recommendation 14: The NSTC Subcommittees on MLAI and NITRD, in conjunction with the NSTC Committee on Science, Technology, Engineering, and Education (CoSTEM),, should initiate a study on the AI workforce pipeline in order to develop actions that ensure an appropriate increase in the size, quality, and diversity of the workforce, including AI researchers, specialists, and users.

Recommendation 15: The Executive Office of the President should publish a follow-on report by the end of this year, to further investigate the effects of AI and automation on the U.S. job market, and outline recommended policy responses.

Recommendation 16: Federal agencies that use AI-based systems to make or provide decision support for consequential decisions about individuals should take extra care to ensure the efficacy and fairness of those systems, based on evidence-based verification and validation.

Recommendation 17: Federal agencies that make grants to state and local governments in support of the use of AI-based systems to make consequential decisions about individuals should review the terms of grants to ensure that AI-based products or services purchased with Federal grant funds produce results in a sufficiently transparent fashion and are supported by evidence of efficacy and fairness.

Recommendation 18: Schools and universities should include ethics, and related topics in security, privacy, and safety, as an integral part of curricula on AI, machine learning, computer science, and data science.

Recommendation 19: AI professionals, safety professionals, and their professional societies should work together to continue progress toward a mature field of AI safety engineering.

Recommendation 20: The U.S. Government should develop a government-wide strategy on international engagement related to AI, and develop a list of AI topical areas that need international engagement and monitoring.

Recommendation 21: The U.S. Government should deepen its engagement with key international stakeholders, including foreign governments, international organizations, industry, academia, and others, to exchange information and facilitate collaboration on AI R&D.

Recommendation 22: Agencies' plans and strategies should account for the influence of AI on cybersecurity, and of cybersecurity on AI. Agencies involved in AI issues should engage their U.S. Government and private-sector cybersecurity colleagues for input on how to ensure that AI systems and ecosystems are secure and resilient to intelligent adversaries. Agencies involved in cybersecurity issues should engage their U.S. Government and private sector AI colleagues for innovative ways to apply AI for effective and efficient cybersecurity.

Recommendation 23: The U.S. Government should complete the development of a single, government-wide policy, consistent with international humanitarian law, on autonomous and semi-autonomous weapons.

Acronyms

AAAI	Association for the Advancement of Artificial Intelligence
AGI	Artificial General Intelligence
AI	Artificial Intelligence
APEC	Asia-Pacific Economic Cooperation
BRAIN	Brain Research through Advancing Innovative Neurotechnologies
CALO	Cognitive Agent that Learns and Organizes
CCC	Computing Community Consortium
CCW	Convention on Certain Conventional Weapons
CEA	Council of Economic Advisers
CEO	Chief Executive Officer
CGC	Cyber Grand Challenge (run by DARPA)
CoSTEM	Committee on Science Technology, Engineering, and Education (component of NSTC)
CS	Computer Science
DARPA	Defense Advanced Research Projects Agency
DoD	Department of Defense
DOT	Department of Transportation
FAA	Federal Aviation Administration
FMVSS	Federal Motor Vehicle Safety Standards
IARPA	Intelligence Advanced Research Projects Activity
ICTs	Information and Communication Technologies
IPA	Intergovernmental Personnel Act
LAWS	Lethal Autonomous Weapon Systems
MLAI	Machine Learning and Artificial Intelligence (subcommittee of NSTC)
NAS	National Airspace System
NEC	National Economic Council
NHTSA	National Highway Traffic Safety Administration
NIH	National Institutes of Health
NIPS	Neural Information Processing Systems conference
NITRD	Networking and Information Technology Research and Development (subcommittee of NSTC)
NSF	National Science Foundation
NSTC	National Science and Technology Council

OECD	Organization for Economic Cooperation and Development
OMB	Office of Management and Budget
ONR	Office of Naval Research
OSTP	Office of Science and Technology Policy
R&D	Research and Development
RFI	Request For Information
STEM	Science, Technology, Engineering, and Mathematics
UAS	Unmanned Aerial System
UTM	UAS Traffic Management

References

"AAAI Presidential Panel on Long-Term AI Futures: 2008-2009 Study," *The Association for the Advancement of Artificial Intelligence*, http://www.aaai.org/Organization/presidential-panel.php.

"Aerospace Forecast Report Fiscal Years 2016 to 2036," *The Federal Aviation Administration*, March 24 016, https://www.faa.gov/data_research/aviation/aerospace_forecasts/media/Unmanned_Aircraft_Systems.pdf.

"AI Safety Conference in Puerto Rico," *The Future of Life Institute,* October 12, 2015, http://futureoflife.org/2015/10/12/ai-safety-conference-in-puerto-rico.

"Big Data: Seizing Opportunities, Preserving Values," *Executive Office of the President*, May 2014, https://www.whitehouse.gov/sites/default/files/docs/big_data_privacy_report_may_1_2014.pdf.

"Data Science for Social Good," *University of Chicago, https://dssg.uchicago.edu/.*

"Federal Automated Vehicles Policy," *The U.S. Department of Transportation*, September 21 2016, https://www.transportation.gov/AV.

"Federal Cybersecurity Research and Development Strategic Plan," *Executive Office of the President,* February 2016, https://www.whitehouse.gov/sites/whitehouse.gov/files/documents/2016_Federal_Cybersecurity_Research_and_Development_Stratgeic_Plan.pdf.

"Secretary Foxx Unveils President Obama's FY17 Budget Proposal of Nearly $4 Billion for Automated Vehicles and Announces DOT Initiatives to Accelerate Vehicle Safety Innovations," *U.S. Department of Transportation*, January 14 2016, https://www.transportation.gov/briefing-room/secretary-foxx-unveils-president-obama%E2%80%99s-fy17-budget-proposal-nearly-4-billion.

"Winning the Education Future: The Role of ARPA-ED," *The U.S. Department of Education*, March 8 2011, https://www.whitehouse.gov/sites/default/files/microsites/ostp/arpa-ed-factsheet.pdf.

"World Development Report 2016: Digital Dividends," *The World Bank Group*, 2016, http://documents.worldbank.org/curated/en/896971468194972881/pdf/102725-PUB-Replacement-PUBLIC.pdf.

Dario Amodei, Chris Olah, Jacob Steinhardt, Paul Christiano, John Schulman, and Dan Mané, "Concrete Problems in AI Safety," https://arxiv.org/abs/1606.06565.

Julia Angwin, Jeff Larson, Surya Mattu, and Lauren Kirchner, "Machine Bias," *ProPublica,* May 23, 2016, https://www.propublica.org/article/machine-bias-risk-assessments-in-criminal-sentencing.

Stuart Armstrong, Kaj Sotala, Seán S. ÓhÉigeartaigh, "The errors, insights and lessons of famous AI predictions – and what they mean for the future," *Journal of Experimental & Theoretical Artificial Intelligence*, May 20, 2014.

Michael Ball, Cynthia Barnhart, Martin Dresner, Mark Hansen, Kevin Neels, Amedeo Odoni, Everett Peterson, Lance Sherry, Antonio Trani, Bo Zou, "Total Delay Impact Study: A Comprehensive Assessment of the Costs and Impacts of Flight Delay in the United States," *The National Center of Excellence for Aviation Operations Research*, November 2010, http://www.nextor.org/pubs/TDI_Report_Final_11_03_10.pdf.

Nicholas Bloom, Mark Schankerman, John Van Reene, "Identifying Technology Spillovers and Product Market Rivalry," *Econometrica*, 81: 1347–1393. doi:10.3982/ECTA9466. Frank Chen, "AI, Deep Learning, and Machine Learning: A Primer," *Andreessen Horowitz*, June 10, 2016, http://a16z.com/2016/06/10/ai-deep-learning-machines.

Jeffrey L. Caton, "Autonomous Weapons Systems: A Brief Survey of Developmental, Operational, Legal, and Ethical Issues," Strategic Studies Institute, U.S. Army War College, December 2015, http://www.strategicstudiesinstitute.army.mil/pdffiles/PUB1309.pdf.

Jack Clark, "Artificial Intelligence Has a 'Sea of Dudes' Problem," *Bloomberg*, June 21, 2016, https://www.bloomberg.com/news/articles/2016-06-23/artificial-intelligence-has-a-sea-of-dudes-problem.

Christianne Corbett and Catherine Hill, "Solving the Equation: The Variables for Women's Success in Engineering and Computing," *The American Association of University Women*, March 2015, http://www.aauw.org/files/2015/03/Solving-the-Equation-report-nsa.pdf.

Anupam Datta, Shayak Sen, and Yair Zick, "Algorithmic Transparency via Quantitative Input Influence: Theory and Experiments with Learning Systems," *Proceedings of 37th IEEE Symposium on Security and Privacy*, 2016.

Pedro Domingos, *The Master Algorithm: How the Quest for the Ultimate Learning Machine Will Remake Our World* (New York, New York: Basic Books, 2015).

Eric Elster, "Surgical Critical Care Initiative: Bringing Precision Medicine to the Critically Ill," presentation at AI for Social Good workshop, Washington, DC, June 7, 2016, http://cra.org/ccc/wp-content/uploads/sites/2/2016/06/Eric-Elster-AI-slides-min.pdf.

Heinz Erzberger, "The Automated Airspace Concept," prepared for the 4th USA/Europe Air Traffic Management R&D Seminar Dec. 3-7, 2001, Santa Fe, New Mexico, USA, http://www.aviationsystemsdivision.arc.nasa.gov/publications/tactical/erzberger_12_01.pdf.

Ed Felten and Terah Lyons, "Public Input and Next Steps on the Future of Artificial Intelligence," *Medium*, September 6 2016, https://medium.com/@USCTO/public-input-and-next-steps-on-the-future-of-artificial-intelligence-458b82059fc3.

J.D. Fletcher, "Digital Tutoring in Information Systems Technology for Veterans: Data Report," *The Institute for Defense Analysis*, September 2014.

Matt Ford, "The Missing Statistics of Criminal Justice," *The Atlantic*, May 31, 2015, http://www.theatlantic.com/politics/archive/2015/05/what-we-dont-know-about-mass-incarceration/394520/

Jason Furman, "Is This Time Different? The Opportunities and Challenges of Artificial Intelligence," (presentation, AI Now: The Social and Economic Implications of Artificial Intelligence Technologies in the Near Term, New York, NY, July 7, 2016), Available at https://www.whitehouse.gov/sites/default/files/page/files/20160707_cea_ai_furman.pdf.

Ian J. Goodfellow, Jonathon Shlens, and Christian Szegedy, "Explaining and Harnessing Adversarial Examples," http://arxiv.org/pdf/1412.6572.pdf.

Georg Graetz and Guy Michaels, "Robots at Work," *CEPR Discussion Paper No. DP10477*, March 2015, http://papers.ssrn.com/sol3/papers.cfm?abstract_id=2575781.

Bronwyn H. Hall, Jacques Mairesse, and Pierre Mohnen, "Measuring the Returns to R&D," Chapter prepared for the Handbook of the Economics of Innovation, B. H. Hall and N. Rosenberg (editors), December 10, 2009, https://eml.berkeley.edu/~bhhall/papers/HallMairesseMohnen09_rndsurvey_HEI.pdf.

Moritz Hardt, "How big data is unfair," *Medium,* September 26 2014, https://medium.com/@mrtz/how-big-data-is-unfair-9aa544d739de.

Neal Jean, Marshall Burke, Michael Xie, W. Matthew Davis, David B. Lobell, and Stefano Ermon. "Combining satellite imagery and machine learning to predict poverty." *Science* 353, no. 6301 (2016): 790-794.

Derryl Jenkins and Bijan Vasigh, "The Economic Impact of Unmanned Aircraft Systems Integration in the United States," *The Association for Unmanned Vehicle Systems International*, 2013, https://higherlogicdownload.s3.amazonaws.com/AUVSI/958c920a-7f9b-4ad2-9807-f9a4e95d1ef1/UploadedImages/New_Economic%20Report%202013%20Full.pdf.

Charles I. Jones and John C. Williams, "Measuring the Social Returns to R&D," *The Quarterly Journal of Economics* (1998) 113 (4): 1119-1135, doi: 10.1162/003355398555856.

Thomas Kalil, "A Broader Vision for Government Research," *Issues in Science and Technology*, 2003.

Garry Kasparov, "The Chess Master and the Computer," *New York Review of Books*, February 11, 2010. http://www.nybooks.com/articles/2010/02/11/the-chess-master-and-the-computer.

Katharine E. Henry, David N. Hager, Peter J. Pronovost, and Suchi Saria, "A targeted real-time early warning score (TREWScore) for septic shock," *Science Translational Medicine* 7, no. 299 (2015): 299ra122-299ra122.

Aimee Leslie, Christine Hof, Diego Amorocho, Tanya Berger-Wolf, Jason Holmberg, Chuck Stewart, Stephen G. Dunbar, and Claire Jea,, "The Internet of Turtles," April 12, 2016, https://www.researchgate.net/publication/301202821_The_Internet_of_Turtles.

Steven Levy, "How Google is Remaking Itself as a Machine Learning First Company," *Backchannel*, June 22, 2016, https://backchannel.com/how-google-is-remaking-itself-as-a-machine-learning-first-company-ada63defcb70.

John Markoff, "No Sailors Needed: Robot Sailboats Scout the Oceans for Data," *The New York Times*, September 4, 2016.

Warren S. McCulloch and Walter H. Pitts, "A Logical Calculus of the Ideas Immanent in Nervous Activity," *Bulletin of Mathematical Biophysics*, 5:115-133, 1943.

Vincent Müller and Nick Bostrom, "Future progress in artificial intelligence: A Survey of Expert Opinion," *Fundamental Issues of Artificial Intelligence*, 2014.

Carrie Mullins, "Retrospective Analysis of Technology Forecasting," *The Tauri Group*, August 13, 2012.Andrew Nusca, "IBM's CEO Thinks Every Digital Business Will Become a Cognitive Computing Business," *Fortune*, June 1 2016.

Robert W. Poole, Jr., "The Urgent Need to Reform the FAA's Air Traffic Control System," *The Heritage Foundation*, 2007, http://www.heritage.org/research/reports/2007/02/the-urgent-need-to-reform-the-faas-air-traffic-control-system.

The President's Council of Advisors on Science and Technology, letter to the President, September 2014, https://www.whitehouse.gov/sites/default/files/microsites/ostp/PCAST/pcast_workforce_edit_report_sept_2014.pdf.

The President's Council of Advisors on Science and Technology, "Report to the President: Big Data and Privacy: A Technological Perspective," *Executive Office of the President*, May 2014, https://www.whitehouse.gov/sites/default/files/microsites/ostp/PCAST/pcast_big_data_and_privacy_-_may_2014.pdf.

Mike Purdy and Paul Daugherty, "Why Artificial Intelligence is the Future of Growth," *Accenture*, 2016, https://www.accenture.com/us-en/_acnmedia/PDF-33/Accenture-Why-AI-is-the-Future-of-Growth.pdf.

Sara Reardon, "Text-mining offers clues to success: US intelligence programme analyses language in patents and papers to identify next big technologies," *Nature* no. 509, 410 (May 22 2014).

David Robinson and Logan Koepke, "Stuck in a Pattern: Early evidence on 'predictive policing' and civil rights," *Upturn*, August 2016, http://www.stuckinapattern.org.

Stuart Russell and Peter Norvig, *Artificial Intelligence: A Modern Approach (3rd Edition)* (Essex, England: Pearson, 2009).

Mary Shaw, Prospects for an Engineering Discipline of Software, IEEE Software 7(6), November 1990.

Stephen F. Smith, "Smart Infrastructure for Urban Mobility," presentation at AI for Social Good workshop, Washington, DC, June 7, 2016, http://cra.org/ccc/wp-content/uploads/sites/2/2016/06/Stephen-Smith-AI-slides.pdf.

Peter Stone, Rodney Brooks, Erik Brynjolfsson, Ryan Calo, Oren Etzioni, Greg Hager, Julia Hirschberg, Shivaram Kalyanakrishnan, Ece Kamar, Sarit Kraus, Kevin Leyton-Brown, David Parkes, William Press, AnnaLee Saxenian,

Julie Shah, Milind Tambe, and Astro Teller, "Artificial Intelligence and Life in 2030," *One Hundred Year Study on Artificial Intelligence: Report of the 2015-2016 Study Panel*, Stanford University, Stanford, CA, September 2016, http://ai100.stanford.edu/2016-report.

Charles Twardy, Robin Hanson, Kathryn Laskey, Tod S. Levitt, Brandon Goldfedder, Adam Siegel, Bruce D'Ambrosio, and Daniel Maxwell, "SciCast: Collective Forecasting of Innovation," *Collective Intelligence,* 2014.

United States Standards Strategy Committee, "United States standards strategy," *New York: American National Standards Institute* (2015), https://share.ansi.org/shared%20documents/Standards%20Activities/NSSC/USSS_Third_edition/ANSI_USSS_2015.pdf.

Dayong Wang, Aditya Khosla, Rishab Gargeya, Humayun Irshad, Andrew H. Beck, "Deep Learning for Identifying Metastatic Breast Cancer," June 18, 2016, https://arxiv.org/pdf/1606.05718v1.pdf.